ESSAYS IN BIOCHEMISTRY

Other recent titles in the *Essays in Biochemistry* series:
Cell Polarity and Cancer: volume 53
edited by A.D. Chalmers and P. Whitley
2012
ISBN 978 1 85578 189 4

Lysine-Based Post-Translational Modification of Proteins: volume 52
edited by I. Scott
2012
ISBN 978 1 85578 185 6

Molecular Parasitology: volume 51
edited by R. Docampo
2011
ISBN 978 1 85578 184 9

ABC Transporters: volume 50
edited by F.J. Sharom
2011
ISBN 978 1 85578 181 8

Chronobiology: volume 49
edited by H.D. Piggins and C. Guilding
2011
ISBN 978 1 85578 180 1

Epigenetics, Disease and Behaviour: volume 48
edited by H.J. Lipps, J. Postberg and D.A. Jackson
2010
ISBN 978 1 85578 179 5

Mitochondrial Function: volume 47
edited by G.C. Brown and M.P. Murphy
2010
ISBN 978 1 85578 178 8

ESSAYS IN BIOCHEMISTRY

volume 54 2013

The Role of Non-Coding RNAs in Biology

Edited by Mark A. Lindsay and Sam Griffiths-Jones

Series Editor
Nigel Hooper (Leeds, U.K.)

Advisory Board
G. Banting (Bristol, U.K.)
E. Blair (Leeds, U.K.)
P. Brookes (Rochester, NY, U.S.A.)
S. Gutteridge (Newark, DE, U.S.A.)
J. Pearson (London, U. K.)
J. Rossjohn (Melbourne, Australia)
E. Shephard (London, U.K.)
J. Tavaré (Bristol, U.K.)
C. Tournier (Manchester, U.K.)

Essays in Biochemistry is published by Portland Press Limited on behalf of the Biochemical Society

Portland Press Limited
Third Floor, Charles Darwin House
12 Roger Street
London WC1N 2JU
U.K.
Tel: +44 (0)20 7685 2410
Fax: +44 (0)20 7685 2469
email: editorial@portlandpress.com
www.portlandpress.com

© The Authors; Journal compilation © 2013 Biochemical Society

All rights reserved. Apart from any fair dealing for the purposes of research or private study, or criticism or review, as permitted under the Copyright, Designs and Patents Act, 1998, this publication may be reproduced, stored or transmitted, in any forms or by any means, only with the prior permission of the publishers, or in the case of reprographic reproduction in accordance with the terms of the licences issued by the Copyright Licensing Agency. Inquiries concerning reproduction outside those terms should be sent to the publishers at the above-mentioned address.

Although, at the time of going to press, the information contained in this publication is believed to be correct, neither the authors nor the editors nor the publisher assumes any responsibility for any errors or omissions herein contained. Opinions expressed in this book are those of the authors and are not necessarily held by the Biochemical Society, the editors or the publisher.

All profits made from the sale of this publication are returned to the Biochemical Society for the promotion of the molecular life sciences.

British Library Cataloguing-in-Publication Data
A catalogue record for this book is available from the British Library
ISBN   978-1-85578-190-0
ISSN (print)   0071 1365
ISSN (online) 1744 1358

Typeset by Techset Composition Ltd, Salisbury, U.K.
Printed in Great Britain by Cambrian Printers Ltd, Aberystwyth

CONTENTS

Preface ix

Authors xi

Abbreviations xv

1 The dark matter rises: the expanding world of regulatory RNAs
Michael B. Clark, Anupma Choudhary, Martin A. Smith, Ryan J. Taft and John S. Mattick

Abstract 1
Introduction 2
The animal genome and the non-coding universe 2
Small ncRNAs 3
lncRNAs (long ncRNAs) 6
Regulatory RNAs in prokaryotes 7
The role of ncRNAs in evolutionary innovation 8
Conclusions 9
Summary 9
References 10

2 Biogenesis and the regulation of the maturation of miRNAs
Nham Tran and Gyorgy Hutvagner

Abstract 17
Introduction 18
Transcription and processing of pri-miRNAs (primary miRNAs) in the nucleus 18
Regulation of pri-miRNA processing 19
Pre-miRNA processing in the cytoplasm 19
Regulation of pre-miRNA processing 20
Loading and activation of miRNAs 20
Regulation of mature miRNA stability 21
Alternative pathways for miRNA biogenesis 21
sno-miRNAs [snoRNA (small nucleolar RNA)-derived miRNAs] 23
Virus-derived miRNAs 24
Conclusion 24
Summary 25
References 25

3 Mechanism of miRNA-mediated repression of mRNA translation
Tamas Dalmay

Abstract	29
Introduction	29
Target recognition	30
Animal miRNAs repress translation	30
Animal miRNAs cause target degradation	31
Plant miRNAs repress translation	33
Plant miRNAs cleave target mRNAs	34
Conclusions	36
Summary	36
References	37

4 Piwi-interacting RNAs: biological functions and biogenesis
Kaoru Sato and Mikiko C. Siomi

Abstract	39
Introduction	40
Targets of piRNA-mediated silencing	40
piRNA biogenesis	41
piRNA pathway factors in *Drosophila*	44
The possible role(s) of piRNAs in cancer	48
Conclusions	49
Summary	49
References	49

5 Small nucleolar RNAs and RNA-guided post-transcriptional modification
Lauren Lui and Todd Lowe

Abstract	53
Introduction	54
Composition and structure of snoRNPs	56
Functions of snoRNAs beyond ribosomal processing and modification	58
Diversity in expression and organization among species	61
Computational search methods and databases	62
snoRNP biogenesis	67
Conclusions	70
Summary	70
References	70

6 Role of small nuclear RNAs in eukaryotic gene expression
Saba Valadkhan and Lalith S. Gunawardane

Abstract	79
Introduction: the challenge of splicing and evolution of eukaryotic snRNAs (small nuclear RNAs)	80
Roles of snRNAs in the spliceosome	82
Beyond splicing: other biological roles of snRNAs	87
Conclusion	87
Summary	88
References	88

7 The functions of natural antisense transcripts
Megan Wight and Andreas Werner

Abstract	91
Introduction	91
How do NATs regulate gene expression?	93
RNA masking	93
A-to-I editing	95
RNAi (RNA interference)	95
TI (transcriptional interference)	96
NAT-induced chromatin changes	96
NATs as drug targets	97
Is there a bigger picture?	97
Conclusions	99
Summary	99
References	100

8 Pseudogenes as regulators of biological function
Ryan C. Pink and David R.F. Carter

Abstract	103
Introduction	103
Transcription	105
Evidence for function	106
Antisense pairing and siRNA (small interfering RNA) production	106
Regulation of mRNA stability	108
Conclusion	109
Summary	109
References	110

9 Identification and function of long non-coding RNAs
Robert S. Young and Chris P. Ponting

Abstract 113
Introduction 113
lincRNA identification 115
Genome-wide indicators of lincRNA functionality 116
lncRNA molecular mechanisms 117
lincRNAs have been implicated in diverse human diseases 121
Conclusions 122
Summary 122
References 123

10 Therapeutic targeting of non-coding RNAs
Thomas C. Roberts and Matthew J.A. Wood

Abstract 127
Introduction 128
miRNAs 128
miRNA therapeutics 128
miRNA antagonism 128
miRNA replacement therapy 132
miRNA-mediated transgene inactivation 132
lncRNAs 132
lncRNA function 133
lncRNA in disease 135
Cross-talk between short and long RNAs 137
Conclusions 138
Summary 140
References 140

Index 147

PREFACE

It has been known for many years that some genes in both eukaryotes and prokaryotes are transcribed to produce RNA, but are not translated into proteins. The advent of high-throughput sequencing technology has led to an explosion in the discovery of these 'non-coding RNA' genes, and functional analysis has indicated their diverse biological roles, including the regulation of transcription and translation, catalysis, chromatin structure, RNA processing and transposon silencing. In this volume of *Essays in Biochemistry*, we have attempted to produce an accessible introduction to this rapidly developing area by inviting leading experts to review some of the most important and interesting families of non-coding RNAs. Non-coding RNAs are often subdivided into 'short' and 'long' families, and we have organized the chapters accordingly.

In Chapter 1, Michael Clark, Anupma Choudhary, Martin Smith, Ryan Taft and John Mattick provide a concise overview of non-coding RNAs in both eukaryotic and prokaryotic organisms. The small non-coding RNAs section begins with two chapters on the most extensively investigated family, the microRNAs. In Chapter 2, Nham Tran and Gyorgy Hutvagner examine the mechanism of microRNA biogenesis, whereas Chapter 3, by Tamas Dalmay, discusses the mechanisms by which microRNAs regulate the translation of protein-coding genes via the RNA interference pathway. Subsequent chapters examine the biogenesis and role of other important families of small RNAs. In Chapter 4, Kauro Sato and Mikiko Siomi assess the importance of piwi-interacting RNAs in maintaining germline integrity through the silencing of transposable elements. In Chapter 5, Lauren Lui and Todd Lowe examine the role of small nucleolar RNAs in guiding the post–transcriptional modification of other non-coding RNAs, as well as recent studies that implicate these RNAs in RNA silencing, telomerase maintenance, alternative splicing and human cancer. The small RNA section is completed by Saba Valadkhan and Lalith Gunawardane, who survey the role of small nuclear RNAs in splicing as well as other aspects of RNA biogenesis, including polyadenylation and RNA stability (Chapter 6).

In general, much less is known about the evolution and function of long non-coding RNAs. These sequences are therefore commonly classified based on their origin and/or genomic context: for example, natural antisense (Chapter 7), pseudogenes (Chapter 8) and long intergenic non-coding RNAs (Chapter 9). The chapter on natural antisense by Megan Wight and Andreas Werner (Chapter 7) considers the evidence that the pairing of antisense with sense transcripts can regulate gene expression, and the potential importance of this mechanism in disease. Although pseudogenes are no longer able to produce proteins, there is increasing evidence that many are transcribed, and that this might have an impact on protein expression from other loci. In Chapter 8, Ryan Pink and David Carter assess the evidence to support this contention, the possible mechanisms and their potential biological function. The section on long non-coding RNAs is finished by Robert Young and Chris Ponting (Chapter 9), who discuss the various approaches that have been employed to identify long intergenic non-coding RNAs and their biological functions. Finally, Chapter 10, by Thomas Roberts and Matthew Woods, considers the potential of targeting non–coding RNAs as a novel therapeutic approach.

We thank the *Essays in Biochemistry* Editorial Advisory Panel and all members of the Portland Press staff for their work on producing this volume. We particularly thank Clare Curtis for her efficient management of the entire process. We also thank the authors for their efforts in producing high-quality reviews in a timely manner, and the anonymous reviewers for their comments and suggestions.

Mark Lindsay and Sam Griffiths-Jones
March 2013

AUTHORS

Mark Lindsay obtained a degree in Natural Sciences from Cambridge University in 1986 and a Ph.D. investigating the mechanism of insulin release from Nottingham in 1991. In the intervening years he has held positions at Imperial College London, AstraZeneca Pharmaceuticals and the University of Manchester. In 2011, he moved to the position of Chair in Molecular Pharmacology at the University of Bath. Dr Lindsay has spent the last 10 years examining the role of non-coding RNAs in inflammation and respiratory diseases such as asthma, chronic obstructive pulmonary disease and cancer.

Sam Griffiths-Jones is a Senior Lecturer in the Faculty of Life Sciences, University of Manchester. Dr Griffiths-Jones has research interests in the computational analysis of non-coding RNA biology. In particular, his group work on the evolution, annotation and function of microRNAs, including management and production of the miRBase database of microRNA sequences. From 2001 to 2006, Dr Griffiths-Jones worked at the Wellcome Trust Sanger Institute, where he founded miRBase and the Rfam database of RNA families, and worked on the Pfam and Interpro protein family resources. Prior to that he obtained a degree in Biochemistry and Biological Chemistry (1997) and a Ph.D. in Chemistry (2001) from the University of Nottingham.

Michael B. Clark is a research officer at the Institute for Molecular Bioscience at the University of Queensland. His research focuses on transcriptomics, long non-coding RNAs and how RNA-based regulation underlies human traits and diseases.

Anupma Choudhary is a Ph.D. student in the Taft Laboratory at the Institute for Molecular Bioscience, University of Queensland. She has a Bachelor of Science with Honours in Microbiology and a Master's degree in Genetics from the University of Delhi, India. Her research interests lie in understanding novel features and functions of small non-coding RNAs, with particular focus on deciphering the regulatory roles of nuclear microRNAs.

Martin Smith completed a B.Sc. in Biological Sciences at the Université de Montréal, where he also obtained an M.Sc. in Bioinformatics in collaboration with the Infectious Disease Research Center of the Centre Hospitalier de l'Université Laval and the McGill Center for Bioinformatics. Dr Smith investigated the impact of a class of short interspersed degenerated retroposons on gene expression and genome organization in the parasitic protozoan *Leishmania*. He then obtained a Ph.D. from the University of Queensland in genomics and computational biology at the Institute for Molecular Bioscience. His Ph.D. thesis is entitled '*Widespread purifying selection on RNA structure in mammals*'. Dr Smith is currently a research officer at the Garvan Institute of Medical Research in the RNA Biology and Plasticity group in Sydney.

Ryan Taft is a Laboratory Head and Australian Research Council Discovery Early Career Research fellow at the Institute for Molecular Bioscience at the

University of Queensland. His work is focused on the cryptic genomics that underlies the RNA regulatory world and rare inherited disease.

John Mattick is head of the RNA Biology and Plasticity Laboratory at the Garvan Institute of Medical Research. He is the executive director of the Garvan Institute, Conjoint Professor in the St Vincent's Clinical School and Visiting Professorial Fellow, School of Biotechnology and Biomolecular Science, University of New South Wales. His work focuses on the role of RNA regulatory systems in the evolution and development of complex multicellular organisms.

Nham Tran obtained his Ph.D. at Johnson & Johnson Research and the University of New South Wales, Australia. During this period, he studied with Dr Greg Arndt in designing a dual-promoter system for expressing small interfering RNAs in mammalian cells and using long double-stranded RNA to regulate genes. In 2005 he joined Professor Christopher O'Brien and Professor Barbara Rose at Sydney University to elucidate the role of small RNAs in head and neck cancer. This group was the first to perform a genome-wide screen of microRNAs in head and neck cancer. In 2010, Dr Tran was awarded the Chancellor's Fellowship at the University of Technology, Sydney, to further his work in small RNAs. During this period, he won the early career development award from the CDMRP–DoD (Congressionally Directed Medical Research Programs of the Department of Defense) to investigate the use of exosomal small RNA as biomarkers for early cancer detection. In 2013 he established his own independent laboratory with the aim of using small RNA biomarkers for the early detection of head and neck cancer. His laboratory is also focused on trying to understand the function of microRNAs in this disease.

Gyorgy Hutvagner obtained his Ph.D. at the Szent Istvan University in Hungary. He won several Hungarian and international short-term fellowships which allowed him to work in Sweden and in The Netherlands. In 2001, he joined Dr Phillip Zamore's laboratory at The University of Massachusetts Medical School as a postdoctoral fellow. As a postdoc, he studied the biochemical pathway of RNA interference and contributed to several key discoveries. In 2002, he won the postdoctoral fellowship of the Medical Foundation. In 2005, Dr Hutvagner joined the Division of Gene Regulation and Expression at the School of Life Sciences at the University of Dundee as a lecturer and an independent investigator. He is also the recipient of the prestigious Wellcome Trust Career Development Fellowship. In 2011 he took up a new post as an Associate Professor at the University of Technology, Sydney, and was awarded the Australian Research Council Future Fellowship. His laboratory continues to unveil the mechanism of microRNA-mediated gene silencing in human cells. Also his laboratory is involved in the identification and characterization of novel small regulatory RNAs.

Tamas Dalmay graduated in Budapest and did his Ph.D. in Molecular Plant Virology in Godollo, Hungary. He moved to Norwich in 1995 with an EMBO fellowship to work with Professor David Baulcombe on the genetics of gene silencing. He obtained a Lectureship in 2002 and became a Professor in 2012 at the University of East Anglia where his group has been working on the biology of small non-coding RNAs in animal and plant systems.

Kaoru Sato studied for a Ph.D. in Life Sciences at the Graduate School of Frontier Sciences, the University of Tokyo, and received his Ph.D. in 2009. He subsequently undertook postdoctoral training with Professor Haruhiko Siomi at Keio University School of Medicine, where he started ongoing studies on the molecular mechanisms of RNA silencing. In 2012, he became an Assistant Professor at Keio University School of Medicine. In 2013, he moved to the University of Tokyo as an Assistant Professor, where he works with Professor Mikiko C. Siomi.

Mikiko C. Siomi is a Professor in the Graduate School of Science in the University of Tokyo, Japan. She was awarded Ph.D.s from Kyoto University in 1994 and the University of Tokushima in 2003. In 1999, together with Haruhiko Siomi, she founded a laboratory at the University of Tokushima to study the molecular function of *Drosophila* FMR1 (dFMR1). Later, her research focused on small-RNA-mediated gene silencing pathways in *Drosophila*. In 2012, she moved to her present position at the University of Tokyo, where her study of small RNA-mediated gene silencing continues.

Lauren Lui is a graduate student in the Biomolecular Engineering and Bioinformatics program at the University of California, Santa Cruz. She received her B.S. from the University of California, Davis, as a Mathematical and Scientific Computation major.

Todd Lowe is an Associate Professor in the Department of Biomolecular Engineering, and member of the Center for Molecular Biology of RNA at the University of California, Santa Cruz.

Saba Valadkhan started her scientific career in Professor James Manley's laboratory in Columbia University as a graduate student, where she worked on elucidating the structure and function of U6 and U2 small nuclear RNAs. In 2004, she joined the Center for RNA Molecular Biology in Case Western Reserve University School of Medicine as an Assistant Professor and continued her work on the catalytic properties of the *in vitro* assembled protein-free complex of U6 and U2 snRNAs and their relationship to spliceosomal catalysis.

Lalith Gunawardane is currently a postdoctoral scientist in the Valadkhan laboratory in the Center for RNA Molecular Biology at Case Western Reserve University. As a graduate student in Professor Haruhiko Siomi's laboratory in Keio University School of Medicine in Tokyo, he worked on the small non-coding RNAs and their interactions with proteins. Since joining the Valadkhan laboratory, he has continued his work on analysis of function of nuclear non-coding RNAs in mammalian cells.

Megan Wight graduated from Newcastle University with an honours degree in Biomedical Sciences specializing in the genetics of complex disease. Her research project focuses on non-coding RNAs with particular emphasis on natural antisense transcripts.

Andreas Werner received his degree in Biochemistry and a Ph.D. in Physiology from the University of Zurich, Switzerland. He is now a Reader for Molecular Physiology at the Institute for Cell and Molecular Biosciences at Newcastle

University, U.K. The serendipitous discovery of a naturally occurring antisense transcript in 1995 prompted his interest in non-protein-coding RNAs. His research focuses on molecular mechanisms of gene regulation by antisense RNAs. The aim is to understand a bigger picture of antisense transcription in the context of evolution and organismal complexity.

Ryan Pink is a postdoctoral research fellow at Oxford Brookes University. He received his Ph.D. from Cranfield University in 2007 working on molecular links of diet and oesophageal cancer in South Africa. This was followed by his first postdoctoral position working on novel cancer nucleotide-based sensors. In 2009, Dr Pink moved to Oxford Brookes University to develop novel methods for RNA analysis in blood systems. He now works on the role of non-coding RNAs in the regulation of both cancer and blood disorders. Dr Pink also has awards for his prolific science engagement and arts–science cross-over projects.

David Carter received his Ph.D. from Cambridge University, U.K., in 2003, where he developed a novel technique to detect the physical interaction between the β-globin gene and its enhancer, the locus control region. He moved to Oxford University to pursue postdoctoral research in the structure of the nucleus and how this influences gene expression. He then spent 2 years as a lecturer at Cranfield University before taking up a senior lectureship at Oxford Brookes University where he spends most of his time getting very excited about non-coding RNAs.

Robert Young completed his D.Phil. entitled 'Evolution and Function of long non-coding RNAs in *Drosophila*' in Chris Ponting's laboratory in 2011. He is currently working as a postdoc at the MRC Human Genetics Unit, investigating transcribed enhancers and their mechanisms of action.

Chris Ponting is Deputy Director of the MRC Functional Genomics Unit, University of Oxford, and Associate Faculty member of the Wellcome Trust Sanger Institute. His ERC Advanced Grant has allowed his group to focus on the computational and experimental characterization of long non-coding RNAs in both vertebrate and invertebrate model systems.

Thomas Roberts is a doctoral candidate working in the Department of Physiology, Anatomy and Genetics at the University of Oxford. His research has focused on small-RNA-mediated epigenetic modulation of therapeutic target genes and the role of differential microRNA expression in the pathophysiology of Duchenne muscular dystrophy. His research interests include RNA biology, non-viral gene therapy and neuromuscular disorders.

Matthew Wood is Professor of Neuroscience at the University of Oxford, and Fellow and Tutor in Medicine at Somerville College, Oxford. He directs a research group investigating RNA biology and the development of RNA-based therapies for neurological and neuromuscular diseases.

ABBREVIATIONS

ABC	ATP-binding cassette
ADAR	adenosine deaminases that act on RNA
aHIF	antisense hypoxia-inducible factor
AMO	anti-miRNA oligonucleotide
ANRIL	CDKN2B antisense RNA 1
APC	antigen-presenting cell
BACE1	β-secretase 1
Bcd1	box C/D RNA 1
BDNF	brain-derived neurotrophic factor
BLV	bovine leukaemia virus
CLL	chronic lymphocytic leukaemia
CPSF	cleavage and polyadenylation stimulating factor
CRC	colorectal cancer
CRISPR	clustered regularly interspersed short palindromic repeat
crRNA	CRISPR RNA
cuff	*cutoff*
DGCR8	Di George Syndrome critical region gene 8
DHFR	dihydrofolate reductase
DMD	Duchenne muscular dystrophy
dsDNA	double-stranded DNA
dsRNA	double-stranded RNA
EBV	Epstein–Barr virus
eIF4G	eukaryotic translation-initiation factor 4G
EM	electron microscopy
eRNA	enhancer RNA
EST	expressed sequence tag
FasIII	*Fasciclin III*
flam	*flamenco*
FS(1)Yb	Female Sterile (1) Yb
GAR	glycine–arginine rich
Gas–5	growth arrest–specific 5
GFP	green fluorescent protein
GR	glucocorticoid receptor
H3K9me3	histone H3 Lys^9 trimethylation
HCV	hepatitis C virus
HDAC1	histone deacetylase 1
HDE	histone downstream element
HMG	high-mobility group
HMGA1	high-mobility group A1
hnRNP	heterogeneous nuclear ribonucleoprotein

hnRNPA1	heterogeneous nuclear ribonucleoprotein A1
HP1	heterochromatin protein 1
Hsp	heat-shock protein
ICG	interchromatin granule
Igf2r	insulin-like growth factor 2 receptor
IRES	internal ribosome entry site
ISL	intramolecular stem-loop
KHSRP	KH-type splicing regulatory protein
KSHV	Kaposi's sarcoma-associated herpes virus
lincRNA	long/large intergenic ncRNA
LNA	locked nucleic acid
lncRNA	long ncRNAs
Mael	Maelstrom
MALAT–1	metastasis associated in lung adenocarcinoma transcript-1
MCPIP1	monocyte chemoattractant protein-1-induced protein 1
MDV	Marek's disease virus
miRISC	miRNA-induced silencing complex
miRLC	miRNA loading complex
miRNA	microRNA
MLE	Maleless
MOF	Males absent on the first
MSL	male-specific lethal
MTOC	microtubule-organizing centre
Naf1	nuclear assembly factor 1
NAT	natural antisense transcript
ncRNA	non-coding RNA
NFAT	nuclear factor of activated T-cells
nos	*nanos*
NSCLC	non-small cell lung cancer
ORF	open reading frame
PABC	cytoplasmic polyA-binding protein
PABP	polyA-binding protein
PAPI	Partner of PIWIs
PARE	parallel analysis of RNA ends
PASR	promoter-associated small RNA
piRISC	piRNA-induced silencing complex
piRNA	piwi-interacting RNA
PNA	peptide nucleic acid
PRC2	Polycomb Repressive Complex 2
pre-miRNA	precursor miRNA
pri-miRNA	primary miRNA
PRMT5	protein arginine N-methyltransferase 5
pRNA	promoter-associated RNA

PROMPT	promoter upstream transcripts
P-TEFb	positive transcription elongation factor
PTEN	phosphatase and tensin homologue deleted on chromosome 10
PUA	pseudouridine and archeosine transglycosylase
PWS	Prader–Willi syndrome
RdRP	RNA-dependent RNA polymerase
Rhi	*Rhino*
RISC	RNA–induced silencing complex
RITS	RNA-induced transcriptional silencing complex
RLC	RISC loading complex
RNA Pol II	RNA polymerase II
RNAa	RNA activation
RNAi	RNA interference
RNP	ribonucleoprotein
rRNA	ribosomal RNA
scaRNP	small Cajal body-specific RNA
sDMA	symmetric dimethylarginine residue
sdRNA	sno-derived RNA
SILAC	stable isotope labelling by amino acids in cell culture
siRNA	small interfering RNA
SL	spliced leader
sno-miRNA	snoRNA (small nucleolar RNA)-derived miRNA
snoRNA	small nucleolar RNA
snRNA	small nuclear RNA
snRNP	small nuclear ribonucleoprotein
SNV	single nucleotide variant
spliRNA	splice site RNA
sRNA	small RNA (prokaryote only)
sRNA	sno-like RNA
Ste	*Stellate*
stRNA	small temporal RNA
Su(Ste)	*Suppressor of Stellate*
TASR	gene termini-associated small RNA
TDRD	Tud domain-containing
TE	transposable element
TERT	telomerase reverse transcriptase
TFIIH	transcription factor II H
TGA	transcriptional gene activation
TGS	transcriptional gene silencing
TI	transcriptional interference
tiRNA	transcription initiation RNA
tj	*traffic jam*
TR	telomerase RNA

TRα2	thyroid hormone receptor α2
TRBP	*trans*-activation response RNA-binding protein
TSSa-RNA	transcription start-site-associated RNAs
TUTase	terminal uridyl transferase
UTR	untranslated region
VEGF	vascular endothelial growth factor
XCI	X chromosome inactivation
Xist	X-inactive specific transcript
Zuc	*Zucchini*

The dark matter rises: the expanding world of regulatory RNAs

Michael B. Clark[*†1], Anupma Choudhary[*], Martin A. Smith[*†], Ryan J. Taft[*] and John S. Mattick[†1]

*Institute for Molecular Bioscience, The University of Queensland, Brisbane, QLD 4072, Australia

†Garvan Institute of Medical Research, Darlinghurst, Sydney, NSW 2010, Australia

Abstract

The ability to sequence genomes and characterize their products has begun to reveal the central role for regulatory RNAs in biology, especially in complex organisms. It is now evident that the human genome contains not only protein-coding genes, but also tens of thousands of non–protein coding genes that express small and long ncRNAs (non-coding RNAs). Rapid progress in characterizing these ncRNAs has identified a diverse range of subclasses, which vary widely in size, sequence and mechanism-of-action, but share a common functional theme of regulating gene expression. ncRNAs play a crucial role in many cellular pathways, including the differentiation and development of cells and organs and, when misregulated, in a number of diseases. Increasing evidence suggests that these RNAs are a major area of evolutionary innovation and play an important role in determining phenotypic diversity in animals.

Keywords:
non-coding RNA, regulatory RNA, regulation of gene expression, small RNA.

[1]Correspondence may be addressed to either of these authors (email m.clark3@uq.edu.au or j.mattick@garvan.org.au).

Introduction

Recent advances in molecular biology, led by the large-scale sequencing of genomes and the characterization of transcriptomes, have revealed that animal genomes are far more complex and intricate than previously anticipated, containing a great diversity of sequences that can be effectors of genetic information, i.e. genes. Central to this has been the discovery that many genes do not encode proteins, but rather produce ncRNAs (non-coding RNAs), sometimes referred to as genomic 'dark matter'. Unlike in prokaryotes and most unicellular eukaryotes, such as yeast, only a small fraction of the mammalian genome encodes protein, yet the vast majority of the mammalian genome is transcribed, producing tens of thousands of small and long ncRNAs [1,2].

The properties of RNA molecules, including their ability to form higher-order structures, to specifically hybridize with other RNAs or DNA, and to assemble RNA–protein complexes, makes them effective and versatile regulatory molecules that can direct relatively generic effector proteins to sequence-specific targets [3]. The functional versatility of RNA has led to the 'RNA world hypothesis', which postulates that the dual catalytic and informational storage properties of RNA provided the molecular platform for the evolution of early life [4]. While proteins and DNA now fulfil most catalytic and information storage roles respectively, RNA continues to have a variety of functional roles, including as a regulator, in all kingdoms of life [4]. The emergence of multicellular life, however, has required increasingly complex regulatory circuits to orchestrate the development and organization of specialized tissues and organs. Hence, although prokaryotes and unicellular eukaryotes do contain regulatory ncRNAs, the role of ncRNAs as regulators of gene expression has seemingly expanded in the genomes of multicellular organisms [5].

The explosion of ncRNA research has revealed that there is an abundance of small and long ncRNAs involved in regulating almost all steps of gene expression, including, but not limited to, chromatin modification, transcriptional control, mRNA degradation, translational efficiency and splicing [6,7]. Hence, ncRNAs function in a wide range of cellular processes, play crucial roles in development and disease, and may even play a central role in the evolution of different species and complex organisms [8–10].

The animal genome and the non-coding universe

For half a century, protein-centric convictions predicated on the central dogma of molecular biology dominated the discipline. However, the completion of multiple genome sequencing projects has revealed that protein-coding sequences encompass only a small fraction of animal genomes and less than 2% of the genome in humans and other mammals [11,12]. Thus either animal genomes are largely composed of 'junk' DNA, or they contain another form of genetic information that has thus far been overlooked. Several lines of evidence support the latter, including the positive correlation between the proportion of the genome that is 'junk'/non-protein-coding and developmental complexity [5], the presence of extensive conserved non-coding sequences outside protein-coding regions [13], and the pervasive differential transcription of the vast majority of the genome [1,14].

Characterization of the mammalian transcriptome has revealed that RNA transcripts are produced from a considerably greater proportion of the genome than the ~40% covered (including introns and exons) by known genes [1,15,16]. This pervasive transcription of the genome, defined as occurring when "the majority of (the genome's) bases are associated with at least one primary transcript" [17] has been identified by a number of independent techniques, including genome-wide tiling arrays, large-scale cloning and sequencing of cDNAs, and next-generation RNA sequencing. For example, results from the ENCODE project, which aimed to identify all functional elements within the human genome, demonstrated that at least 75% of bases were transcribed [1]. A number of studies have shown that although the vast majority of the genome is transcribed at some level, most transcription, including novel unannotated transcription, clusters around known genes [16,18–20]. Such analyses led Kapranov et al. [19] to propose "a model of genome organization where protein–coding genes are at the center of a complex network of overlapping sense and antisense (long) RNA transcription, with interleaved (small) RNAs" (Figure 1).

The net result of pervasive transcription is a complex and interleaved transcriptome producing approximately 20000 coding genes, along with at least as many, and possibly a much greater number of, transcripts that do not encode proteins and could function at the RNA level [2]. These can be separated into two major groups (the small and long ncRNAs) on the basis of size and mechanism of synthesis.

Small ncRNAs

Small RNAs are generally defined as ncRNAs shorter than 200 nt in length, and are usually produced by the post–transcriptional processing of longer transcripts by endogenous

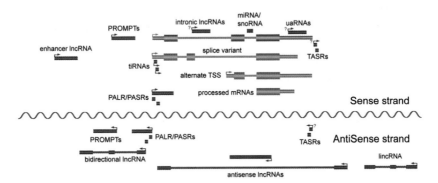

Figure 1. Pervasive transcription around a hypothetical protein coding gene
A standard mRNA transcript is shown in blue, other coding or potentially coding transcripts are shown in green, and non-coding transcripts are in red. Introns are represented by thin lines, non-coding exons are indicated by medium thickness lines and coding exons are indicated by the thickest lines. Arrows represent transcription start sites. The absence of an arrow indicates that the transcripts are generated by processing. An arrow plus a question mark refers to transcripts where the origin is often unclear, and could involve transcription initiation or processing. PALR, promoter-associated long RNA [19]; PASR, promoter-associated small RNA [19]; PROMPT, promoter upstream transcript [51]; snoRNA, small nucleolar RNA; TASR, termini-associated small RNA [19]; tiRNA, transcription initiation (tiny) RNA [49]; TSS, transcription start site; uaRNA, 3′-UTR-derived RNAs [118].

RNases (i.e. RNA cleavage enzymes). Based primarily on their size, mode of biogenesis and function, small RNAs can be divided into various subclasses. The three subclasses of small RNAs that have thus far received the most attention are those that participate in RNAi (RNA interference) pathways, namely, miRNAs (microRNAs), siRNAs (small interfering RNAs) and piRNAs (piwi-interacting RNAs), which produce mature RNAs ~20–30 nt in length. Nucleotide sequence complementarity lies at the heart of the widespread and potent regulatory control that these small RNAs exert on their targets. In all known RNAi pathways, this regulatory control is mediated by binding of the small RNA to a complex of proteins, chief among them being the Argonaute proteins. There are also slightly larger small RNA species (~100 nt) that have important cellular roles, including snoRNAs (small nucleolar RNAs), which guide RNA base modification [21] [and can also be processed into other classes of regulatory small RNA, including miRNAs and sdRNAs (sno-derived RNAs)] [22]; snRNAs (small nuclear RNAs), which mediate RNA splicing and are important components of the spliceosome [23]; Y RNAs, which appear to regulate the Ro autoantigen [24]; and vault RNAs, components of the vault ribonucleoprotein complex [25]. Most small ncRNAs function as part of RNA–protein complexes to regulate gene expression, with the small RNA often acting to specify the target for regulation through nucleotide sequence complementarity (Figures 2A and 2B) [26].

miRNAs are ~22 nt long and bind to short regions of complementary sequence, usually located in the 3′ UTR (untranslated region), of target mRNAs [27]. miRNAs can bring about the translational repression or degradation of target transcripts, depending on the extent of complementarity between them (Figure 2A) [26]. The ability of these small RNAs to modulate gene expression was first identified 20 years ago with the discovery of *lin-4* in the nematode worm *Caenorhabditis elegans* [28,29]. Since then, the miRNA field has evolved rapidly and there are now over 1500 miRNAs annotated in the human genome [30]. Mature miRNAs are produced as part of a two-step enzymatic process from a longer pri-miRNA (primary miRNA) that is processed into a pre-miRNA (precursor miRNA) in the nucleus. The pre-miRNA is then transported to the cytoplasm where it is cleaved to release mature miRNAs [31]. The mature miRNA is then loaded on to the RISC (RNA-induced silencing complex), which is composed of the Argonaute2 protein, the miRNA and other auxiliary proteins [26]. The significance of miRNAs in biological processes is highlighted by the fact that one miRNA can potentially target, and hence control the expression of, many hundreds of mRNAs [32,33]. miRNAs can also regulate transcription through mechanisms that are not yet fully understood in mammalian cells [34]. However, it seems likely that there is more to miRNA function than just the inhibition of translation. There are increasing reports describing the presence of mature miRNAs in the nucleus [35,36], suggesting that they may also be directly or indirectly involved in transcriptional gene silencing.

siRNAs are ~21 nt long and function mainly by degrading transcripts they have perfect sequence complementarity to. The precursor transcripts of endogenous siRNAs include dsRNAs (double-stranded RNAs), pseudogenes [37] or transcripts with very long stem-loop structures [38,39]. Endogenous siRNAs have been proposed to protect eukaryotic cells from dsRNA viruses and are also important for silencing transposons and other 'selfish' genomic elements [40,41]. Since siRNAs use the same enzymatic machinery as miRNAs to function, synthetic siRNAs can be introduced into cells to 'knock-down' any given gene, a feature that has been exploited in scientific research and in a number of recently developed therapeutics [42].

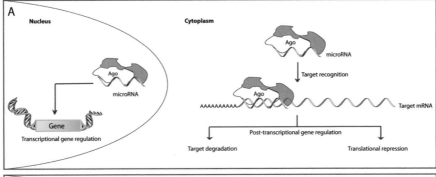
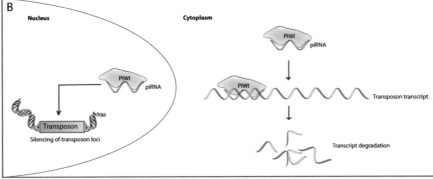
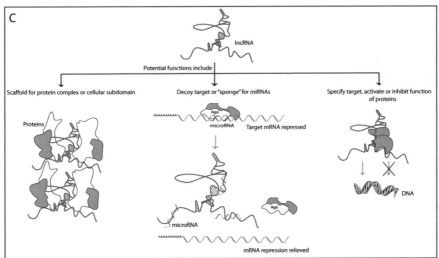

Figure 2. Examples of ncRNA function
(**A**) Transcriptional and translational regulation by miRNAs (shown in blue), which are expressed in nearly every tissue and cell type in complex animals, most simple animals, plants and fungi. Mature miRNAs are loaded into an Argonaute (Ago) protein-containing RISC complex in the cytoplasm. Argonaute proteins have also been reported to function in the nucleus of various organisms. (**B**) piRNA silencing of transposons in germ cells. piRNAs (shown in blue and white) bind to PIWI proteins (shown in green) in the germ line and cause the degradation of transposon transcripts. (**C**) Some potential functions of a lncRNA. Folded lncRNA is shown in red. Proteins are shown in white or grey. miRNAs are in blue. The series of A on the target mRNAs represents the polyA tail. DNA is shown as a double helix.

piRNAs are a related class of small (28–32 nt) RNAs that are found only in animals and principally expressed in spermatids [43]. piRNAs are found in clusters in the genome, and appear to arise from long single-stranded precursor transcripts [44,45]. Unlike miRNAs and siRNAs, which bind to the Ago subclade of the Argonaute proteins, piRNAs associate with the Piwi subclass of Argonaute proteins. They have an important role in suppressing the expression of repetitive elements by guiding DNA methylation, and have been shown to be involved in gametogenesis (Figure 2B) [46,47], as well as in neuronal plasticity in the sea hare, *Aplysia* [48].

Apart from these major categories, other small ncRNAs have been described that originate from regions adjacent to transcription start-sites, such as tiRNAs (transcription initiation RNAs). These RNAs are ~17–18 nt long and are abundant at active promoters as well as at loci with evidence of bidirectional transcription [49], and have been shown to influence the epigenetic state of the genomic region from which they are derived [50]. RNAs of similar size [spliRNAs (splice site RNAs)] are also associated with splice sites [36]. Other, but less well-defined, small RNAs found at or near transcription start-sites that have been reported include PASRs (promoter-associated small RNAs) [19], PROMPTs (promoter upstream transcripts) [51], TSSa-RNAs (transcription start-site-associated RNAs) [52] and TASRs (gene termini-associated small RNAs) [19].

Consistent with their role in gene regulation, small RNAs are involved in many cellular processes and their dysfunction is implicated in a number of physiological and developmental defects [53]. For example, aberrant expression of miRNAs is implicated in a wide variety of diseases ranging from disorders of the heart to immune diseases and cancer [54]. Additionally, a snoRNA has been demonstrated to play a central role in Prader–Willi syndrome [55], whereas another has been associated with autism [56].

lncRNAs (long ncRNAs)

lncRNAs are generally defined as ranging in size from ~200 nt to over 100 kb in length [57,58]. Although the 200 nt cut-off for lncRNAs is quite arbitrary, it has the advantage of excluding most transcripts accepted to be members of small RNA classes. Unlike small ncRNAs, lncRNAs cannot be easily divided into different subclasses on the basis of sequence characteristics and mode-of-action, and this inability to classify lncRNAs into different subtypes has contributed to their current arbitrary definition. We have previously suggested a more flexible definition that lncRNAs are "noncoding RNAs that may have a function as either primary or spliced transcripts, which are independent of processing into known classes of small RNAs, such as miRNAs, piRNAs and snoRNAs, while also excluding structural RNAs from classical housekeeping families" [59], such as rRNAs.

The presence of lncRNAs with important functions has been known for some time, with the characterization of lncRNAs such as *Xist* (X-inactive specific transcript, which controls the silencing of one X chromosome in female mammals) in the 1990s [60,61]. However, the first database of eukaryotic lncRNAs had less than 12 entries by the end of the millennium [62]. The identification of thousands of putative lncRNA transcripts from genome-wide transcriptome analysis [15,63,64], along with prominent examples of functional lncRNAs [65–67], demonstrated that lncRNAs such as *Xist* were not rare genomic oddities, but were instead the first characterized examples of a large class of novel genes. By the end of 2010 over 100 lncRNAs had been functionally characterized as part of a surge of research into the lncRNA world [59]. The subset of lncRNAs transcribed from intergenic regions (i.e. genomic loci some distance from and not overlapping protein-coding genes) have received the most recent attention and are known as lincRNAs (long/large intergenic ncRNAs). Along with those from intergenic loci, lncRNAs are transcribed from many other regions of the genome including promoters, enhancers, introns, UTRs, as overlapping or non-coding isoforms of coding genes, antisense to other transcripts and from pseudogenes [15,68–70].

Although often of similar length to mRNAs, there are a number of differences between mRNAs and lncRNAs beyond the absence of a functional ORF (open reading frame) in the latter. Analysis of lncRNA expression has shown they have lower expression levels and are more likely to be expressed in highly tissue- and cell-specific patterns [63,71–73]. Unlike most mRNAs and many small ncRNAs, lncRNAs are not as highly conserved, although they do show evidence of conservation in their promoters, primary sequences and splice sites [15,71,73–75]. Furthermore, many lncRNAs consist of a single exon and those that are spliced have fewer exons than protein-coding genes [63,72,73]. lncRNAs also commonly contain transposable elements and other repeats [73]. Sequences from such genetic elements can be 'domesticated' during evolution and contribute to lncRNA function by promoting their expression and providing functional motifs [76].

lncRNAs carry out a diverse range of functions in the cell (Figure 2C). Although few are reported to function catalytically, many carry out RNA–protein, RNA–DNA and RNA–RNA interactions. Similar to many small ncRNAs, lncRNAs can regulate gene expression via RNA–protein (ribonucleoprotein) complexes (Figure 2C) [77]. A common function of lncRNAs appears to be directing the activity of chromatin-modifying complexes and transcription factors by specifying their genomic DNA targets and activating or inhibiting their function [67,78–83]. In these and other contexts, lncRNAs have the ability to act as scaffolds, nucleating the assembly of larger complexes or cellular structures [84–86]. Other reported lncRNA functions include acting as miRNA sponges to 'soak up' miRNAs, relieving the repression of mRNAs and so controlling mRNA expression levels and mRNA translation [87,88].

lncRNAs can function both locally and *in trans*. An example of the former is *Airn*, which silences the expression of neighbouring genes to regulate imprinting of the *Igf2r* (insulin-like growth factor 2 receptor) locus [57,65]. *Trans*-acting lncRNAs include *HOTAIR*, which is expressed from the *HOXC* locus and acts to silence gene expression at many genomic locations, including the *HOXD* locus, by recruiting repressive chromatin modification complexes [67,89].

Given their range of functions, a number of lncRNAs are also implicated or involved in disease states, including functioning as oncogenes or tumour suppressors [87,90,91], as well as being linked to other complex diseases such as myocardial infarction [92] and Alzheimer's disease [93].

Regulatory RNAs in prokaryotes

The versatility of RNA as a regulator has also been used by prokaryotes, which contain numerous ncRNAs. Similar to eukaryotes, prokaryotic ncRNAs are being found to play an increasingly important role in regulating gene expression [94]. Despite these similarities, few ncRNA classes are shared between prokaryotes and eukaryotes, with the exception of snoRNAs, which are present in archaea (although not in bacteria) [95].

Many prokaryotic sRNAs (small RNAs; abbreviation only used in prokaryotes) and antisense RNAs function as ncRNAs. sRNAs are generally defined as transcripts <500 nt and can be expressed from any region of the genome. Antisense RNAs include both sRNAs and longer RNAs that are antisense to coding genes, creating some overlap between the sRNA and antisense RNA classes. ncRNAs are common in prokaryotic genomes, with approximately 170 non-coding sRNAs predicted in the archaea *Methanosarcina mazei* [96], ~50 in the bacteria *Listeria monocytogenes* [97] and 165 in *Pseudomonas aeruginosa* [98]. In comparison, *Pseudomonas* was found to contain 384 antisense RNAs [98], whereas *Helicobacter pylori* was reported to have antisense transcription across most of the genome, covering 46% of ORFs [99].

Prokaryotic regulatory RNAs function by a variety of mechanisms (reviewed in [94,100]). *Cis*-antisense ncRNAs commonly act to repress the expression of the sense coding gene. Repression can occur either at the level of transcription, RNA turnover or by inhibiting translation, with the extensive nucleotide complementarity between the two important for many of these mechanisms. Examples include ncRNAs regulating the copy number of mobile elements such as plasmids and repressing the translation of toxic proteins, such as the SymR antisense sRNA in *Escherichia coli* that represses the synthesis of the SymE toxin protein [101,102].

Trans-encoded sRNAs generally act to repress translation or destabilize target RNA(s), demonstrating some functional similarity to eukaryotic miRNAs. With more limited complementarity than *cis*-antisense RNAs, *trans* RNAs often require the RNA chaperone protein Hfq to bind their targets [94]. An example is the association of four sRNAs with Hfq in *Vibrio cholerae* to control quorum sensing by destabilizing the mRNA of the quorum-sensing master regulator [103]. Some sRNAs also bind directly to proteins by mimicking other nucleic acid sequences. For example, the 6S sRNA mimics the structure of an open promoter to bind RNA polymerase and regulate transcription [94,104].

Another important class of prokaryotic ncRNAs are CRISPRs (clustered regularly interspersed short palindromic repeats). First discovered in *E. coli* [105], CRISPRs are now known to exist in most bacteria and archaea [106]. CRISPR loci contain short direct repeats interspersed with spacer regions derived from invading mobile elements. CRISPRs are transcribed and processed to generate small crRNAs (CRISPR RNAs), which function to protect the cell from invading bacteriophages and conjugative plasmids [106,107].

The role of ncRNAs in evolutionary innovation

The sequencing and initial annotation of mammalian genomes provided two large surprises: the large fraction of the genome comprised of sequences derived from transposable elements and the much lower than expected number of protein-coding genes [11,12]. In fact, the number of recognized human protein-coding genes (20687) [2] is similar to that in the nematode worm *C. elegans* (20517) [108] and that in a basal metazoan, the sponge *Amphimedon queenslandica* (18500–30000) [109]. Furthermore, much of the protein-coding 'toolkit' that controls multicellular processes in more complex animals is also present in the sponge [109]. There is widespread use of alternative splicing in human genes [110], which can diversify the proteome without an increase in gene number, suggesting that it is one mechanism to explain differences in complexity. However, alternative splicing itself requires regulation and hence it has been hypothesized that increases in gene regulatory complexity underlie much of morphological complexity [5,8,10].

Moreover, given the relatively stable protein-coding complement, it is clear that most evolutionary adaptation occurs in regulatory sequences, which are fast evolving and show little conservation over long evolutionary distances [111–113]. The discovery that most of this non-coding DNA is dynamically transcribed to generate tens of thousands of ncRNAs [1,15,16] provides a hitherto unexpected mechanism to explain this increase in regulatory complexity.

ncRNAs, with their potential to bind DNA, RNA and protein in a sequence- or structure-specific manner, are versatile and effective regulatory molecules. By providing specificity to generic protein complexes [3], ncRNAs can act as guides to selectively target effector proteins to different loci and thereby regulate the transcription or translation of many genes [90,114].

Lastly the pervasive transcription of different ncRNAs in the genome provides a large dynamic pool of transcripts for selection to act upon, as most ncRNAs are subject to more flexible structure–function constraints than protein-coding RNAs [17,111,115]. For instance, many ncRNAs function via the formation of stable secondary and tertiary structures, which can accommodate compensatory nucleotide substitutions, e.g. A:U base pairs to G:U or G:C, without disrupting their structural (and thus functional) integrity [116]. Moreover, regulatory sequences are also subject to positive selection for adaptive radiation [117]. The overarching conclusion is that regulatory ncRNAs represent a vast hidden layer of evolutionarily plastic *cis*- and *trans*-acting regulatory information that directs the epigenetic pathways that underpin animal development and diversity [8–10].

Conclusions

The last decade has revolutionized our understanding of genomes and what constitutes a gene. It has become increasingly apparent that many cellular functions are mediated by RNA, a realization that has far-reaching implications for understanding human biology and treating human disease. The following chapters outline the state-of-the-art in the characterization of various types of ncRNAs, although the continuing rapid pace of discovery and unknown function of so many ncRNAs makes it clear that much remains to be done before this poorly charted sphere of biology is fully explored.

Summary
- Non-coding RNA genes are abundant in the genome, with similar numbers of protein-coding and non-coding genes in humans.
- Non-coding RNAs are structurally diverse, ranging from less than 20 nt to over 100 kb in length.
- The properties of RNA molecules allow them to function through both sequence complementarity to other RNAs or DNA, as well as forming structures that can interact with proteins and/or nucleic acids.

- Many well-characterized subclasses of small non-coding RNAs are known, with each member of a subclass having a similar functional mechanism, whereas the subclasses and mechanism-of-action of long non-coding RNAs are much less well understood.
- Most functionally characterized non-coding RNAs (whether small or long) function in the regulation of gene expression.
- Non-coding RNAs play essential roles in many biological processes and are crucial for development and disease, and perhaps even the evolution of organisms.

The authors acknowledge the support of the Australian NHMRC (National Health and Medical Research Council) [NHMRC Australia Fellowship number 631668 (to J.S.M.)], the Australian Research Council [DECRA Fellowship (to R.J.T.)] and the University of Queensland [University of Queensland International Research Tuition Award and University of Queensland Research Scholarship (to A.C.)].

References

1. Djebali, S., Davis, C.A., Merkel, A., Dobin, A., Lassmann, T., Mortazavi, A., Tanzer, A., Lagarde, J., Lin, W., Schlesinger, F. et al. (2012) Landscape of transcription in human cells. Nature **489**, 101–108
2. Harrow, J., Frankish, A., Gonzalez, J.M., Tapanari, E., Diekhans, M., Kokocinski, F., Aken, B.L., Barrell, D., Zadissa, A., Searle, S. et al. (2012) GENCODE: the reference human genome annotation for The ENCODE Project. Genome Res. **22**, 1760–1774
3. Huttenhofer, A. and Schattner, P. (2006) The principles of guiding by RNA: chimeric RNA-protein enzymes. Nat. Rev. Genet. **7**, 475–482
4. Atkins, J.F., Gesteland, R.F. and Cech, T. (2011) RNA worlds: from life's origins to diversity in gene regulation. Cold Spring Harbor Laboratory Press, Cold Spring Harbor
5. Taft, R.J., Pheasant, M. and Mattick, J.S. (2007) The relationship between non-protein-coding DNA and eukaryotic complexity. BioEssays **29**, 288–299
6. Amaral, P.P., Dinger, M.E., Mercer, T.R. and Mattick, J.S. (2008) The eukaryotic genome as an RNA machine. Science **319**, 1787–1789
7. Brosnan, C.A. and Voinnet, O. (2009) The long and the short of noncoding RNAs. Curr. Opin. Cell Biol. **21**, 416–425
8. Mattick, J.S. and Makunin, I.V. (2006) Non-coding RNA. Hum. Mol. Genet. **15**, R17–R29
9. Prasanth, K.V. and Spector, D.L. (2007) Eukaryotic regulatory RNAs: an answer to the 'genome complexity' conundrum. Genes Dev. **21**, 11–42
10. Mattick, J.S. (2011) The central role of RNA in human development and cognition. FEBS Lett. **585**, 1600–1616
11. Lander, E.S., Linton, L.M., Birren, B., Nusbaum, C., Zody, M.C., Baldwin, J., Devon, K., Dewar, K., Doyle, M., FitzHugh, W. et al. (2001) Initial sequencing and analysis of the human genome. Nature **409**, 860–921
12. Waterston, R.H., Lindblad-Toh, K., Birney, E., Rogers, J., Abril, J.F., Agarwal, P., Agarwala, R., Ainscough, R., Alexandersson, M., An, P. et al. (2002) Initial sequencing and comparative analysis of the mouse genome. Nature **420**, 520–562
13. Stephen, S., Pheasant, M., Makunin, I.V. and Mattick, J.S. (2008) Large-scale appearance of ultraconserved elements in tetrapod genomes and slowdown of the molecular clock. Mol. Biol. Evol. **25**, 402–408

14. Clark, M.B., Amaral, P.P., Schlesinger, F.J., Dinger, M.E., Taft, R.J., Rinn, J.L., Ponting, C.P., Stadler, P.F., Morris, K.V., Morillon, A. et al. (2011) The reality of pervasive transcription. PLoS Biol. **9**, e1000625
15. Carninci, P., Kasukawa, T., Katayama, S., Gough, J., Frith, M.C., Maeda, N., Oyama, R., Ravasi, T., Lenhard, B., Wells, C. et al. (2005) The transcriptional landscape of the mammalian genome. Science **309**, 1559–1563
16. Cheng, J., Kapranov, P., Drenkow, J., Dike, S., Brubaker, S., Patel, S., Long, J., Stern, D., Tammana, H., Helt, G. et al. (2005) Transcriptional maps of 10 human chromosomes at 5-nucleotide resolution. Science **308**, 1149–1154
17. Birney, E., Stamatoyannopoulos, J.A., Dutta, A., Guigo, R., Gingeras, T.R., Margulies, E.H., Weng, Z., Snyder, M., Dermitzakis, E.T., Thurman, R.E. et al. (2007) Identification and analysis of functional elements in 1% of the human genome by the ENCODE pilot project. Nature **447**, 799–816
18. Kampa, D., Cheng, J., Kapranov, P., Yamanaka, M., Brubaker, S., Cawley, S., Drenkow, J., Piccolboni, A., Bekiranov, S., Helt, G. et al. (2004) Novel RNAs identified from an in-depth analysis of the transcriptome of human chromosomes 21 and 22. Genome Res. **14**, 331–342
19. Kapranov, P., Cheng, J., Dike, S., Nix, D.A., Duttagupta, R., Willingham, A.T., Stadler, P.F., Hertel, J., Hackermuller, J., Hofacker, I.L. et al. (2007) RNA maps reveal new RNA classes and a possible function for pervasive transcription. Science **316**, 1484–1488
20. Fejes-Toth, K., Sotirova, V., Sachidanandam, R., Assaf, G., Hannon, G.J., Kapranov, P., Foissac, S., Willingham, A.T., Duttagupta, R., Dumais, R. and Gingeras, T.R. (2009) Post-transcriptional processing generates a diversity of 5′-modified long and short RNAs. Nature **457**, 1028–1032
21. Bachellerie, J.P., Cavaille, J. and Huttenhofer, A. (2002) The expanding snoRNA world. Biochimie **84**, 775–790
22. Taft, R.J., Glazov, E.A., Lassmann, T., Hayashizaki, Y., Carninci, P. and Mattick, J.S. (2009) Small RNAs derived from snoRNAs. RNA **15**, 1233–1240
23. Wachtel, C. and Manley, J.L. (2009) Splicing of mRNA precursors: the role of RNAs and proteins in catalysis. Mol. Biosyst. **5**, 311–316
24. Sim, S., Weinberg, D.E., Fuchs, G., Choi, K., Chung, J. and Wolin, S.L. (2009) The subcellular distribution of an RNA quality control protein, the Ro autoantigen, is regulated by noncoding Y RNA binding. Mol. Biol. Cell **20**, 1555–1564
25. Berger, W., Steiner, E., Grusch, M., Elbling, L. and Micksche, M. (2009) Vaults and the major vault protein: novel roles in signal pathway regulation and immunity. Cell. Mol. Life Sci. **66**, 43–61
26. Ghildiyal, M. and Zamore, P.D. (2009) Small silencing RNAs: an expanding universe. Nat. Rev. Genet. **10**, 94–108
27. Lai, E.C. (2002) Micro RNAs are complementary to 3′ UTR sequence motifs that mediate negative post-transcriptional regulation. Nat. Genet. **30**, 363–364
28. Wightman, B., Ha, I. and Ruvkun, G. (1993) Posttranscriptional regulation of the heterochronic gene lin-14 by lin-4 mediates temporal pattern formation in *C. elegans*. Cell **75**, 855–862
29. Lee, R.C., Feinbaum, R.L. and Ambros, V. (1993) The *C. elegans* heterochronic gene lin-4 encodes small RNAs with antisense complementarity to lin-14. Cell **75**, 843–854
30. Kozomara, A. and Griffiths-Jones, S. (2011) miRBase: integrating microRNA annotation and deep-sequencing data. Nucleic Acids Res **39**, D152–D157
31. Lee, Y., Jeon, K., Lee, J.T., Kim, S. and Kim, V.N. (2002) MicroRNA maturation: stepwise processing and subcellular localization. EMBO J. **21**, 4663–4670
32. Landgraf, P., Rusu, M., Sheridan, R., Sewer, A., Iovino, N., Aravin, A., Pfeffer, S., Rice, A., Kamphorst, A.O., Landthaler, M. et al. (2007) A mammalian microRNA expression atlas based on small RNA library sequencing. Cell **129**, 1401–1414
33. Friedman, R.C., Farh, K.K., Burge, C.B. and Bartel, D.P. (2009) Most mammalian mRNAs are conserved targets of microRNAs. Genome Res. **19**, 92–105
34. Khraiwesh, B., Arif, M.A., Seumel, G.I., Ossowski, S., Weigel, D., Reski, R. and Frank, W. (2010) Transcriptional control of gene expression by microRNAs. Cell **140**, 111–122

35. Hwang, H.W., Wentzel, E.A. and Mendell, J.T. (2007) A hexanucleotide element directs microRNA nuclear import. Science **315**, 97–100
36. Taft, R.J., Simons, C., Nahkuri, S., Oey, H., Korbie, D.J., Mercer, T.R., Holst, J., Ritchie, W., Wong, J.J., Rasko, J.E. et al. (2010) Nuclear-localized tiny RNAs are associated with transcription initiation and splice sites in metazoans. Nat. Struct. Mol. Biol. **17**, 1030–1034
37. Watanabe, T., Totoki, Y., Toyoda, A., Kaneda, M., Kuramochi-Miyagawa, S., Obata, Y., Chiba, H., Kohara, Y., Kono, T., Nakano, T. et al. (2008) Endogenous siRNAs from naturally formed dsRNAs regulate transcripts in mouse oocytes. Nature **453**, 539–543
38. Czech, B., Malone, C.D., Zhou, R., Stark, A., Schlingeheyde, C., Dus, M., Perrimon, N., Kellis, M., Wohlschlegel, J.A., Sachidanandam, R. et al. (2008) An endogenous small interfering RNA pathway in *Drosophila*. Nature **453**, 798–802
39. Okamura, K., Balla, S., Martin, R., Liu, N. and Lai, E.C. (2008) Two distinct mechanisms generate endogenous siRNAs from bidirectional transcription in *Drosophila melanogaster*. Nat. Struct. Mol. Biol. **15**, 581–590
40. Kim, V.N., Han, J. and Siomi, M.C. (2009) Biogenesis of small RNAs in animals. Nat. Rev. Mol. Cell Biol. **10**, 126–139
41. Chung, W.J., Okamura, K., Martin, R. and Lai, E.C. (2008) Endogenous RNA interference provides a somatic defense against *Drosophila* transposons. Curr. Biol. **18**, 795–802
42. Li, S.D., Chono, S. and Huang, L. (2008) Efficient oncogene silencing and metastasis inhibition via systemic delivery of siRNA. Mol. Ther. **16**, 942–946
43. Aravin, A., Gaidatzis, D., Pfeffer, S., Lagos-Quintana, M., Landgraf, P., Iovino, N., Morris, P., Brownstein, M.J., Kuramochi-Miyagawa, S., Nakano, T. et al. (2006) A novel class of small RNAs bind to MILI protein in mouse testes. Nature **442**, 203–207
44. Thomson, T. and Lin, H. (2009) The biogenesis and function of PIWI proteins and piRNAs: progress and prospect. Annu. Rev. Cell Dev. Biol. **25**, 355–376
45. Brennecke, J., Aravin, A.A., Stark, A., Dus, M., Kellis, M., Sachidanandam, R. and Hannon, G.J. (2007) Discrete small RNA-generating loci as master regulators of transposon activity in *Drosophila*. Cell **128**, 1089–1103
46. Houwing, S., Kamminga, L.M., Berezikov, E., Cronembold, D., Girard, A., van den Elst, H., Filippov, D.V., Blaser, H., Raz, E., Moens, C.B. et al. (2007) A role for Piwi and piRNAs in germ cell maintenance and transposon silencing in Zebrafish. Cell **129**, 69–82
47. Aravin, A.A., Sachidanandam, R., Bourc'his, D., Schaefer, C., Pezic, D., Toth, K.F., Bestor, T. and Hannon, G.J. (2008) A piRNA pathway primed by individual transposons is linked to *de novo* DNA methylation in mice. Mol. Cell **31**, 785–799
48. Rajasethupathy, P., Antonov, I., Sheridan, R., Frey, S., Sander, C., Tuschl, T. and Kandel, E.R. (2012) A role for neuronal piRNAs in the epigenetic control of memory-related synaptic plasticity. Cell **149**, 693–707
49. Taft, R.J., Glazov, E.A., Cloonan, N., Simons, C., Stephen, S., Faulkner, G.J., Lassmann, T., Forrest, A.R., Grimmond, S.M., Schroder, K. et al. (2009) Tiny RNAs associated with transcription start sites in animals. Nat. Genet. **41**, 572–578
50. Taft, R.J., Hawkins, P.G., Mattick, J.S. and Morris, K.V. (2011) The relationship between transcription initiation RNAs and CCCTC-binding factor (CTCF) localization. Epigenetics Chromatin **4**, 13
51. Preker, P., Nielsen, J., Kammler, S., Lykke-Andersen, S., Christensen, M.S., Mapendano, C.K., Schierup, M.H. and Jensen, T.H. (2008) RNA exosome depletion reveals transcription upstream of active human promoters. Science **322**, 1851–1854
52. Seila, A.C., Calabrese, J.M., Levine, S.S., Yeo, G.W., Rahl, P.B., Flynn, R.A., Young, R.A. and Sharp, P.A. (2008) Divergent transcription from active promoters. Science **322**, 1849–1851
53. Esteller, M. (2011) Non-coding RNAs in human disease. Nat. Rev. Genet. **12**, 861–874
54. Boyd, S.D. (2008) Everything you wanted to know about small RNA but were afraid to ask. Lab. Invest. **88**, 569–578

55. Sahoo, T., del Gaudio, D., German, J.R., Shinawi, M., Peters, S.U., Person, R.E., Garnica, A., Cheung, S.W. and Beaudet, A.L. (2008) Prader–Willi phenotype caused by paternal deficiency for the HBII-85 C/D box small nucleolar RNA cluster. Nat. Genet. **40**, 719–721
56. Nakatani, J., Tamada, K., Hatanaka, F., Ise, S., Ohta, H., Inoue, K., Tomonaga, S., Watanabe, Y., Chung, Y.J., Banerjee, R. et al. (2009) Abnormal behavior in a chromosome-engineered mouse model for human 15q11-13 duplication seen in autism. Cell **137**, 1235–1246
57. Lyle, R., Watanabe, D., te Vruchte, D., Lerchner, W., Smrzka, O.W., Wutz, A., Schageman, J., Hahner, L., Davies, C. and Barlow, D.P. (2000) The imprinted antisense RNA at the Igf2r locus overlaps but does not imprint Mas1. Nat. Genet. **25**, 19–21
58. Furuno, M., Pang, K.C., Ninomiya, N., Fukuda, S., Frith, M.C., Bult, C., Kai, C., Kawai, J., Carninci, P., Hayashizaki, Y. et al. (2006) Clusters of internally primed transcripts reveal novel long noncoding RNAs. PLoS Genet. **2**, e37
59. Amaral, P.P., Clark, M.B., Gascoigne, D.K., Dinger, M.E. and Mattick, J.S. (2011) lncRNAdb: a reference database for long noncoding RNAs. Nucleic Acids Res. **39**, D146–D151
60. Brown, C.J., Hendrich, B.D., Rupert, J.L., Lafreniere, R.G., Xing, Y., Lawrence, J. and Willard, H.F. (1992) The human XIST gene: analysis of a 17 kb inactive X-specific RNA that contains conserved repeats and is highly localized within the nucleus. Cell **71**, 527–542
61. Penny, G.D., Kay, G.F., Sheardown, S.A., Rastan, S. and Brockdorff, N. (1996) Requirement for Xist in X chromosome inactivation. Nature **379**, 131–137
62. Erdmann, V.A., Szymanski, M., Hochberg, A., de Groot, N. and Barciszewski, J. (1999) Collection of mRNA-like non-coding RNAs. Nucleic Acids Res. **27**, 192–195
63. Ravasi, T., Suzuki, H., Pang, K.C., Katayama, S., Furuno, M., Okunishi, R., Fukuda, S., Ru, K., Frith, M.C., Gongora, M.M. et al. (2006) Experimental validation of the regulated expression of large numbers of non-coding RNAs from the mouse genome. Genome Res **16**, 11–19
64. Guttman, M., Amit, I., Garber, M., French, C., Lin, M.F., Feldser, D., Huarte, M., Zuk, O., Carey, B.W., Cassady, J.P. et al. (2009) Chromatin signature reveals over a thousand highly conserved large non-coding RNAs in mammals. Nature **458**, 223–227
65. Sleutels, F., Zwart, R. and Barlow, D.P. (2002) The non-coding Air RNA is required for silencing autosomal imprinted genes. Nature **415**, 810–813
66. Ji, P., Diederichs, S., Wang, W., Boing, S., Metzger, R., Schneider, P.M., Tidow, N., Brandt, B., Buerger, H., Bulk, E. et al. (2003) MALAT-1, a novel noncoding RNA, and thymosin β4 predict metastasis and survival in early–stage non–small cell lung cancer. Oncogene **22**, 8031–8041
67. Rinn, J.L., Kertesz, M., Wang, J.K., Squazzo, S.L., Xu, X., Brugmann, S.A., Goodnough, L.H., Helms, J.A., Farnham, P.J., Segal, E. and Chang, H.Y. (2007) Functional demarcation of active and silent chromatin domains in human HOX loci by noncoding RNAs. Cell **129**, 1311–1323
68. Engstrom, P.G., Suzuki, H., Ninomiya, N., Akalin, A., Sessa, L., Lavorgna, G., Brozzi, A., Luzi, L., Tan, S.L., Yang, L. et al. (2006) Complex loci in human and mouse genomes. PLoS Genet. **2**, e47
69. Nakaya, H.I., Amaral, P.P., Louro, R., Lopes, A., Fachel, A.A., Moreira, Y.B., El-Jundi, T.A., da Silva, A.M., Reis, E.M. and Verjovski-Almeida, S. (2007) Genome mapping and expression analyses of human intronic noncoding RNAs reveal tissue-specific patterns and enrichment in genes related to regulation of transcription. Genome Biol. **8**, R43
70. Kim, T.K., Hemberg, M., Gray, J.M., Costa, A.M., Bear, D.M., Wu, J., Harmin, D.A., Laptewicz, M., Barbara-Haley, K., Kuersten, S. et al. (2010) Widespread transcription at neuronal activity-regulated enhancers. Nature **465**, 182–187
71. Guttman, M., Garber, M., Levin, J.Z., Donaghey, J., Robinson, J., Adiconis, X., Fan, L., Koziol, M.J., Gnirke, A., Nusbaum, C. et al. (2010) *Ab initio* reconstruction of cell type-specific transcriptomes in mouse reveals the conserved multi-exonic structure of lincRNAs. Nat. Biotechnol. **28**, 503–510

72. Cabili, M.N., Trapnell, C., Goff, L., Koziol, M., Tazon-Vega, B., Regev, A. and Rinn, J.L. (2011) Integrative annotation of human large intergenic noncoding RNAs reveals global properties and specific subclasses. Genes Dev. **25**, 1915–1927
73. Derrien, T., Johnson, R., Bussotti, G., Tanzer, A., Djebali, S., Tilgner, H., Guernec, G., Martin, D., Merkel, A., Knowles, D.G. et al. (2012) The GENCODE v7 catalog of human long noncoding RNAs: analysis of their gene structure, evolution, and expression. Genome Res. **22**, 1775–1789
74. Marques, A.C. and Ponting, C.P. (2009) Catalogues of mammalian long noncoding RNAs: modest conservation and incompleteness. Genome Biol. **10**, R124
75. Chodroff, R.A., Goodstadt, L., Sirey, T.M., Oliver, P.L., Davies, K.E., Green, E.D., Molnar, Z. and Ponting, C.P. (2010) Long noncoding RNA genes: conservation of sequence and brain expression among diverse amniotes. Genome Biol. **11**, R72
76. Dinger, M.E., Amaral, P.P., Mercer, T.R. and Mattick, J.S. (2009) Pervasive transcription of the eukaryotic genome: functional indices and conceptual implications. Briefings Funct. Genomics Proteomics **8**, 407–423
77. Rinn, J.L. and Chang, H.Y. (2012) Genome regulation by long noncoding RNAs. Annu. Rev. Biochem. **81**, 145–166
78. Nagano, T., Mitchell, J.A., Sanz, L.A., Pauler, F.M., Ferguson–Smith, A.C., Feil, R. and Fraser, P. (2008) The Air noncoding RNA epigenetically silences transcription by targeting G9a to chromatin. Science **322**, 1717–1720
79. Zhao, J., Sun, B.K., Erwin, J.A., Song, J.J. and Lee, J.T. (2008) Polycomb proteins targeted by a short repeat RNA to the mouse X chromosome. Science **322**, 750–756
80. Huarte, M., Guttman, M., Feldser, D., Garber, M., Koziol, M.J., Kenzelmann-Broz, D., Khalil, A.M., Zuk, O., Amit, I., Rabani, M. et al. (2010) A large intergenic noncoding RNA induced by p53 mediates global gene repression in the p53 response. Cell **142**, 409–419
81. Kino, T., Hurt, D.E., Ichijo, T., Nader, N. and Chrousos, G.P. (2010) Noncoding RNA gas5 is a growth arrest- and starvation-associated repressor of the glucocorticoid receptor. Sci. Signaling **3**, ra8
82. Guttman, M., Donaghey, J., Carey, B.W., Garber, M., Grenier, J.K., Munson, G., Young, G., Lucas, A.B., Ach, R., Bruhn, L. et al. (2011) lincRNAs act in the circuitry controlling pluripotency and differentiation. Nature **477**, 295–300
83. Lanz, R.B., McKenna, N.J., Onate, S.A., Albrecht, U., Wong, J., Tsai, S.Y., Tsai, M.J. and O'Malley, B.W. (1999) A steroid receptor coactivator, SRA, functions as an RNA and is present in an SRC–1 complex. Cell **97**, 17–27
84. Clemson, C.M., Hutchinson, J.N., Sara, S.A., Ensminger, A.W., Fox, A.H., Chess, A. and Lawrence, J.B. (2009) An architectural role for a nuclear noncoding RNA: NEAT1 RNA is essential for the structure of paraspeckles. Mol. Cell **33**, 717–726
85. Sunwoo, H., Dinger, M.E., Wilusz, J.E., Amaral, P.P., Mattick, J.S. and Spector, D.L. (2009) MENε/β nuclear-retained non-coding RNAs are up-regulated upon muscle differentiation and are essential components of paraspeckles. Genome Res. **19**, 347–359
86. Shevtsov, S.P. and Dundr, M. (2011) Nucleation of nuclear bodies by RNA. Nat. Cell Biol. **13**, 167–173
87. Poliseno, L., Salmena, L., Zhang, J., Carver, B., Haveman, W.J. and Pandolfi, P.P. (2010) A coding-independent function of gene and pseudogene mRNAs regulates tumour biology. Nature **465**, 1033–1038
88. Cesana, M., Cacchiarelli, D., Legnini, I., Santini, T., Sthandier, O., Chinappi, M., Tramontano, A. and Bozzoni, I. (2011) A long noncoding RNA controls muscle differentiation by functioning as a competing endogenous RNA. Cell **147**, 358–369
89. Tsai, M.C., Manor, O., Wan, Y., Mosammaparast, N., Wang, J.K., Lan, F., Shi, Y., Segal, E. and Chang, H.Y. (2010) Long noncoding RNA as modular scaffold of histone modification complexes. Science **329**, 689–693
90. Gupta, R.A., Shah, N., Wang, K.C., Kim, J., Horlings, H.M., Wong, D.J., Tsai, M.C., Hung, T., Argani, P., Rinn, J.L. et al. (2010) Long non-coding RNA HOTAIR reprograms chromatin state to promote cancer metastasis. Nature **464**, 1071–1076

91. Zhang, X., Gejman, R., Mahta, A., Zhong, Y., Rice, K.A., Zhou, Y., Cheunsuchon, P., Louis, D.N. and Klibanski, A. (2010) Maternally expressed gene 3, an imprinted noncoding RNA gene, is associated with meningioma pathogenesis and progression. Cancer Res. **70**, 2350–2358
92. Ishii, N., Ozaki, K., Sato, H., Mizuno, H., Saito, S., Takahashi, A., Miyamoto, Y., Ikegawa, S., Kamatani, N., Hori, M. et al. (2006) Identification of a novel non-coding RNA, MIAT, that confers risk of myocardial infarction. J. Hum. Genet. **51**, 1087–1099
93. Mus, E., Hof, P.R. and Tiedge, H. (2007) Dendritic BC200 RNA in aging and in Alzheimer's disease. Proc. Natl. Acad. Sci. U.S.A. **104**, 10679–10684
94. Waters, L.S. and Storz, G. (2009) Regulatory RNAs in bacteria. Cell **136**, 615–628
95. Omer, A.D., Lowe, T.M., Russell, A.G., Ebhardt, H., Eddy, S.R. and Dennis, P.P. (2000) Homologs of small nucleolar RNAs in Archaea. Science **288**, 517–522
96. Jager, D., Sharma, C.M., Thomsen, J., Ehlers, C., Vogel, J. and Schmitz, R.A. (2009) Deep sequencing analysis of the *Methanosarcina mazei* Go1 transcriptome in response to nitrogen availability. Proc. Natl. Acad. Sci. U.S.A. **106**, 21878–21882
97. Toledo-Arana, A., Dussurget, O., Nikitas, G., Sesto, N., Guet-Revillet, H., Balestrino, D., Loh, E., Gripenland, J., Tiensuu, T., Vaitkevicius, K. et al. (2009) The *Listeria* transcriptional landscape from saprophytism to virulence. Nature **459**, 950–956
98. Wurtzel, O., Yoder-Himes, D.R., Han, K., Dandekar, A.A., Edelheit, S., Greenberg, E.P., Sorek, R. and Lory, S. (2012) The single-nucleotide resolution transcriptome of *Pseudomonas aeruginosa* grown in body temperature. PLoS Pathog. **8**, e1002945
99. Sharma, C.M., Hoffmann, S., Darfeuille, F., Reignier, J., Findeiss, S., Sittka, A., Chabas, S., Reiche, K., Hackermuller, J., Reinhardt, R. et al. (2010) The primary transcriptome of the major human pathogen *Helicobacter pylori*. Nature **464**, 250–255
100. Thomason, M.K. and Storz, G. (2010) Bacterial antisense RNAs: how many are there, and what are they doing? Annu. Rev. Genet. **44**, 167–188
101. Tomizawa, J. and Itoh, T. (1981) Plasmid ColE1 incompatibility determined by interaction of RNA I with primer transcript. Proc. Natl. Acad. Sci. U.S.A. **78**, 6096–6100
102. Kawano, M., Aravind, L. and Storz, G. (2007) An antisense RNA controls synthesis of an SOS-induced toxin evolved from an antitoxin. Mol. Microbiol. **64**, 738–754
103. Lenz, D.H., Mok, K.C., Lilley, B.N., Kulkarni, R.V., Wingreen, N.S. and Bassler, B.L. (2004) The small RNA chaperone Hfq and multiple small RNAs control quorum sensing in *Vibrio harveyi* and *Vibrio cholerae*. Cell **118**, 69–82
104. Wassarman, K.M. and Storz, G. (2000) 6S RNA regulates *E. coli* RNA polymerase activity. Cell **101**, 613–623
105. Ishino, Y., Shinagawa, H., Makino, K., Amemura, M. and Nakata, A. (1987) Nucleotide sequence of the iap gene, responsible for alkaline phosphatase isozyme conversion in *Escherichia coli*, and identification of the gene product. J. Bacteriol. **169**, 5429–5433
106. Jore, M.M., Brouns, S.J. and van der Oost, J. (2012) RNA in defense: CRISPRs protect prokaryotes against mobile genetic elements. Cold Spring Harbor Perspect. Biol. **4**, a003657
107. Barrangou, R., Fremaux, C., Deveau, H., Richards, M., Boyaval, P., Moineau, S., Romero, D.A. and Horvath, P. (2007) CRISPR provides acquired resistance against viruses in prokaryotes. Science **315**, 1709–1712
108. Flicek, P., Amode, M.R., Barrell, D., Beal, K., Brent, S., Chen, Y., Clapham, P., Coates, G., Fairley, S., Fitzgerald, S. et al. (2011) Ensembl 2011. Nucleic Acids Res. **39**, D800-D806
109. Srivastava, M., Simakov, O., Chapman, J., Fahey, B., Gauthier, M.E., Mitros, T., Richards, G.S., Conaco, C., Dacre, M., Hellsten, U. et al. (2010) The *Amphimedon queenslandica* genome and the evolution of animal complexity. Nature **466**, 720–726
110. Wang, E.T., Sandberg, R., Luo, S., Khrebtukova, I., Zhang, L., Mayr, C., Kingsmore, S.F., Schroth, G.P. and Burge, C.B. (2008) Alternative isoform regulation in human tissue transcriptomes. Nature **456**, 470–476
111. Meader, S., Ponting, C.P. and Lunter, G. (2010) Massive turnover of functional sequence in human and other mammalian genomes. Genome Res. **20**, 1335–1343

112. Kutter, C., Watt, S., Stefflova, K., Wilson, M.D., Goncalves, A., Ponting, C.P., Odom, D.T. and Marques, A.C. (2012) Rapid turnover of long noncoding RNAs and the evolution of gene expression. PLoS Genet. **8**, e1002841
113. Schmidt, D., Wilson, M.D., Ballester, B., Schwalie, P.C., Brown, G.D., Marshall, A., Kutter, C., Watt, S., Martinez-Jimenez, C.P., Mackay, S. et al. (2010) Five-vertebrate ChIP-seq reveals the evolutionary dynamics of transcription factor binding. Science **328**, 1036–1040
114. Guo, H., Ingolia, N.T., Weissman, J.S. and Bartel, D.P. (2010) Mammalian microRNAs predominantly act to decrease target mRNA levels. Nature **466**, 835–840
115. Heinen, T.J., Staubach, F., Haming, D. and Tautz, D. (2009) Emergence of a new gene from an intergenic region. Curr. Biol. **19**, 1527–1531
116. Smit, S., Knight, R. and Heringa, J. (2009) RNA structure prediction from evolutionary patterns of nucleotide composition. Nucleic Acids Res. **37**, 1378–1386
117. Heimberg, A.M., Sempere, L.F., Moy, V.N., Donoghue, P.C. and Peterson, K.J. (2008) MicroRNAs and the advent of vertebrate morphological complexity. Proc. Natl. Acad. Sci. U.S.A. **105**, 2946–2950
118. Mercer, T.R., Wilhelm, D., Dinger, M.E., Solda, G., Korbie, D.J., Glazov, E.A., Truong, V., Schwenke, M., Simons, C., Matthaei, K.I. et al. (2011) Expression of distinct RNAs from 3′ untranslated regions. Nucleic Acids Res. **39**, 2393–2403

Biogenesis and the regulation of the maturation of miRNAs

Nham Tran*[1] and Gyorgy Hutvagner†[1]

*University of Technology Sydney, Faculty of Science, Centre for Health Technologies, corner of Harris and Thomas Street, Ultimo, Sydney, NSW 2007, Australia
†University of Technology Sydney, Faculty of Engineering and Information Technology, Centre for Health Technologies, 235 Jones Street, Ultimo, Sydney, NSW 2007, Australia

Abstract

Regulation of gene expression is a fundamental process in both prokaryotic and eukaryotic organisms. Multiple regulatory mechanisms are in place to control gene expression at the level of transcription, post-transcription and post-translation to maintain optimal RNA and protein expressions in cells. miRNAs (microRNAs) are abundant short 21–23 nt non-coding RNAs that are key regulators of virtually all eukaryotic biological processes. The levels of miRNAs in an organism are crucial for proper development and sustaining optimal cell functions. Therefore the processing and regulation of the processing of these miRNAs are critical. In the present chapter we highlight the most important steps of miRNA processing, describe the functions of key proteins involved in the maturation of miRNAs, and discuss how the generation and the stability of miRNAs are regulated.

Keywords:
Argonaute, Dicer, Drosha, microRNA.

[1]Correspondence may be addressed to either author (email Nham.Tran@uts.edu.au or gyorgy.hutvagner@uts.edu.au).

Introduction

Historically we had come to understand that gene expression was controlled by transcription factors and other protein co-factors. However, in the last decade it has become more apparent that gene regulation is highly complex due to the discovery of ncRNAs (non-coding RNAs) as key factors in regulating gene expression (Chapter 1). Approximately 15 years ago a new pathway was starting to be elucidated which underpins gene expression in animals and plants. This regulatory pathway is mediated by small ncRNAs, known as miRNAs (microRNAs; also abbreviated to miRs). The first descriptions of these miRNAs were termed stRNAs (small temporal RNAs) and were implicated in *Caenorhabditis elegans* development [1]. In subsequent studies these small RNAs were shown to be common and abundant in plants and animals [2–4], but do not exist in the bacterial kingdom. They are now proven to be the key regulators of gene expression in virtually all biological processes.

The current model for the biogenesis of miRNAs involves several distinct steps. In the nucleus, there is the initial transcription of the miRNA loci and then processing by the RNase III enzyme Drosha. This cleaved product is then transported into the cytoplasm for a second processing event mediated by another RNase III enzyme known as Dicer. The end result of this enzymatic processing is the production of a 21–22 nt dsRNA (double-stranded RNA) species which represents the mature miRNA. The mature miRNA is then loaded on to the Argonaute family of proteins to form the RISC (RNA-induced silencing complex) which is the effector for mediating miRNA activity.

In the present chapter we describe the key steps of miRNA processing and formation of the regulatory competent complex, and also summarize the regulation of many of these steps. We discuss the studies carried out in animal systems and describe the crucial differences between the plant and the animal miRNA pathways.

Transcription and processing of pri-miRNAs (primary miRNAs) in the nucleus

The majority of miRNAs are transcribed by RNA Pol II (RNA polymerase II) which generates a typical RNA transcript containing a 5′ cap and a polyA tail [5]. However, in special cases, RNA Pol III has also been shown to transcribe miRNAs from viral genomes [6]. These RNA polymerases make longer transcripts known as the pri-miRNAs. miRNAs are embedded in the hairpin-structured motifs of these pri-miRNAs. miRNAs are encoded by ncRNAs or introns of protein-coding transcripts. In the initial nuclear processing step for animals, these stem-loops are recognized by the microprocessor complex [7] consisting of Drosha (an RNase III enzyme) and DGCR8 (Di George Syndrome critical region gene 8) [8]. DGCR8 binds to the stem-loop base and then positions Drosha which cleaves the double-stranded stem-loop approximately 11 bp from a less-structured single-stranded base to generate a 100–70 nt hairpin structure known as a pre-miRNA (precursor miRNA) [9]. Plants, however, do not have a Drosha homologue but they have multiple Dicer genes (see below) and one of them generates pre-miRNAs [10]. One of the unique features of Drosha processing is that it generates RNA ends containing a 5′ phosphate group and a 2 nt 3′ overhang, which are characteristic of RNase III activity and are essential for the recognition of the pre-miRNA by further processing enzymes. Apart from its major role in miRNA biogenesis Drosha also recognizes pri-miRNA-like hairpins in protein-coding genes such as Brd2, Mbnl1, Wipi2 and DCGRG, cleaving them at these structures to regulate their expression [11,12]. Other factors known to associate with the microprocessor are the DEAD box RNA helicases p72 and p68, which are needed to process specific subsets of pri-miRNAs [13].

Regulation of pri-miRNA processing

Since miRNAs are key regulators of developmental processes, the cell cycle and immune responses, their processing is also tightly regulated at the different steps of biogenesis. There are an increasing number of mechanisms and proteins that are implicated in the regulation of processing for the pri-miRNAs. First, Drosha and DGCR8 can also regulate each other post-transcriptionally via cleaving a hairpin sequence found on the *DGCR8* transcript, thereby controlling the activity of the microprocessor [12]. It has been shown that transcription factors such as the Smads can bind to the microprocessor via the p68 subunit and enhance Drosha cleavage [14]. In a similar manner the p53 tumour suppressor protein binds to p68 and increases processing for specific pri-miRNAs such as miR-145 [15]. A similar mechanism has been described with BRCA1, another tumour suppressor protein, which facilitates the processing of a subset of pri-miRNAs by directly binding to the pri-mRNAs Drosha and p68. It also interacts with p53 and Smad3, two proteins that regulate signalling and miRNA processing [14]. Hormone-mediated regulation can also occur via the oestrogen receptor which associates with the p68 and p72 helicases [16]. It may be that these RNA helicases act as the scaffold for recruiting factors to enhance the processing of specific pri-miRNAs.

RNA-binding proteins such as KHSRP (KH-type splicing regulatory protein) can also regulate pri-miRNA processing of a subset of miRNAs [17]. It binds with a high affinity to single-stranded GGG-triplet motifs within the terminal loop and promotes microprocessor cleavage of, for instance, the *let-7a* primary transcript. The stimulatory activity of KHSRP can be abrogated by another RNA-binding protein, hnRNPA1 (heterogeneous nuclear ribonucleoprotein A1), which competes for the same binding site at the terminal loop [18]. hnRNPA1 has a dual function since it can also promote pri-miR-18 processing in the miR-17-92 cluster [19]. Lastly, ADARs (adenosine deaminases that act on RNA) are RNA-editing enzymes that deaminate adenosines to create inosines in dsRNA structures. These ADARs can edit a pri-miRNA transcript sequence which then can prevent Drosha processing [20]. These examples clearly suggest a tightly controlled mechanism and also indicate that certain subsets of miRNA families are only processed by binding of specific co-factors to the microprocessor unit. After nuclear processing the next phase occurs in the cytoplasm, with the pre-miRNA being exported in a Ran-GTP-dependent manner through the nuclear export receptor exportin-5 [21]. This transport is specific as exportin-5 recognizes the 3′ overhang and the duplex nature of the pre-miRNA [22].

Pre-miRNA processing in the cytoplasm

In the cytosol, the pre-miRNA is recognized by the second RNase III enzyme Dicer [23] and its partner, the RNA-binding protein [TRBP (*trans*-activation response RNA-binding protein) and PACT in humans/R2D2 and Loquacious in flies, for instance] [24–27]. Dicers in general contain an N-terminal helicase domain, a PAZ domain, two RNase III domains and a C-terminal dsRNA-binding domain. Dicer binds the 3′ overhang and also the 5′ end to correctly position the pre-miRNA for cleavage by its PAZ domain and cleave each of the strands with 2 nt offset with the two RNase III catalytic centres. This process liberates an intermediate product which is a 21–23 nt dsRNA which contains 3′ overhangs at both ends [28]. Recent biochemical and structural studies also suggest that in higher eukaryotes the N-terminal helicase domain is responsible for binding the pre–miRNA loop [29].

Regulation of pre-miRNA processing

Dicer processing of certain pre-miRNAs has been shown to be influenced by the developmentally regulated protein lin-28 in *C. elegans* and mammals [30,31]. This protein has a high specificity for the terminal loop of the let-7 miRNA family and, when bound to the precursor, recruits TUTases (terminal uridyl transferases). These enzymes mediate uridylation of precursor let-7 at its 3′-end which then prevents Dicer processing and promotes degradation of the pre-miRNA [32]. Lin-28 can also be translocated to the nucleus and binds the terminal loop of let-7 primary transcripts to block processing by either physically impeding Drosha binding or sequestering the primary structure into nucleoli [33]. In a twist, Dicer is also regulated by let-7 and therefore there is a negative-feedback mechanism which regulates the stoichiometry of Dicer and its product [34]. Furthermore, MCPIP1 (monocyte chemoattractant protein-1-induced protein 1) is an endo-RNase that regulates miRNA levels by cleaving the loop sequences of a group of miRNAs. This regulation is likely to be important in cancer and immune responses [35].

Loading and activation of miRNAs

Regulatory competent miRNAs are incorporated into members of the Argonaute multiprotein family and form the RISC or miRISC (miRNA-induced silencing complex). Argonaute proteins are structural homologues of RNase H enzymes [36] and characteristically contain an N-terminal domain that is involved in the unwinding of double-stranded small RNAs [37], a PAZ domain (similar to the Dicer PAZ domain with similar function), an MID domain that orientates the miRNA by accommodating the 5′ phosphate of the miRNA and a PIWI motif which is the catalytic domain [38]. All plant Argonautes are cleavage-competent [39]; however, many animal Argonautes have lost this activity and are now involved in cleavage-independent gene regulatory mechanism(s) (Chapter 3). All regulatory small RNAs, including miRNAs, are single-stranded when they regulate gene expression, but are generally loaded on to Argonaute proteins as double-stranded molecules; therefore a mechanism must exist to unwind the double-stranded Dicer product and eliminate one of the two strands. The strand selection of regulatory miRNAs is not random. One strand of the miRNA (the guide or trigger strand) is preferentially loaded on to an Argonaute protein compared with the other strand [the (*) star sequence or passenger strand] of the same miRNA duplex. The strand whose 5′-end is located at the thermodynamically weaker end of the duplex will be the guide strand and it will recognize and bind to cognate RNA targets [40,41].

The RLC (RISC loading complex) and miRLC (miRNA loading complex), in general, contain Dicer, RNA-binding proteins (R2D2, Loquacious in flies and TRBP in mammals) and an Argonaute protein [42]. In the case of siRNAs (small interfering RNAs; similar regulatory RNAs to miRNAs, however they are perfectly double-stranded and processed from mainly long dsRNAs), Dicer and the RNA-binding protein recognize the asymmetry of the siRNA and load them on to a specific Argonaute [43]. However, this sensing has not been verified in the case of miRNAs. For instance, the existence of miRLC has been demonstrated in mammals [44], but it was also shown that Dicer is not required for the recognition of the asymmetry of the miRNA [45].

Recombinant Argonaute can only bind single-stranded RNAs efficiently [46]. The Hsp (heat-shock protein) 70/90 chaperone machinery facilitates the transfer of the Dicer-cleaved double-stranded product to the Argonaute protein by helping to form a conformation that is capable of accommodating double-stranded small RNAs [47]. The unwinding of double-stranded miRNAs is an energy-dependent step whereby the two strands of the duplex are wedged open by the N-terminal domain of the Argonautes. This duplex is then unwound to remove the passenger strand [37]. The single-stranded guide-RNA-activated Argonautes are now primed and ready to engage in diverse gene regulatory functions. A summary of these steps is shown in Figure 1.

Regulation of mature miRNA stability

In general, miRNAs and siRNAs are very stable in the cellular environment with a half-life of several days. However, several miRNAs in specific cells and/or in a defined stage of the cell cycle can show accelerated turnover. Sequence requirements for rapid degradation have been mapped in some miRNAs. Also, RNA-modifying enzymes that tag miRNAs for degradation and nucleases that turn over mature miRNAs have been identified and characterized in plants and animals. The turnover rate of miRNAs can also be regulated by the target mRNA. This regulation depends on the level of the complementarity of an miRNA and its target; more extensive base pairing leads to the degradation of the miRNA through a mechanism called tailing and trimming (mechanisms involved in miRNA turnover are reviewed in detail by Ruegger and Grosshans [48]).

Alternative pathways for miRNA biogenesis

In both *C. elegans* and *Drosophila melanogaster*, there are miRNAs embedded in introns of mRNA-encoding genes which are generated via the action of RNA splicing and lariat-debranching enzymes to produce pre-miRNA-like hairpins known as mirtrons (Figure 2) [49]. These RNA hairpin structures are not processed by Drosha, but are structural mimics resembling pre-miRNAs. These mirtrons then enter the canonical miRNA pathway during nuclear export and are then cleaved by Dicer to generate a mature miRNA. In mammalian cells, mirtrons were found to be well-conserved and expressed in diverse mammals [50]. The discovery of a Drosha-independent processing pathway suggests that mirtrons were the first set of miRNA genes to have evolved for mammalian gene regulation. Remarkably, there are even some mirtrons such as miR-1225 and miR-1228 which are produced in the absence of splicing.

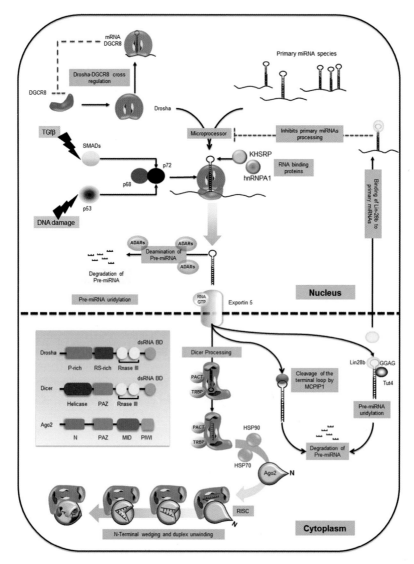

Figure 1. Canonical processing of miRNAs and the regulation of miRNA processing in animal cells

These are termed simtrons and they do not require DGCR8, Dicer, exportin-5 or Ago2 for maturation [51].

Another miRNA which does not require Dicer processing is miR-451. The primary transcript of miR-451 is cleaved via the microprocessor to liberate a precursor which is too small to act as a Dicer substrate, but instead is loaded into Ago2 and is cleaved by the Ago2 catalytic centre [52]. In this instance it is the pre–miRNA secondary structure which determines the processing pathway. Recent studies have indicated that it is a combination of the terminal loop, but more importantly, the nucleotide identity at the 5′-end which sorts the miR-451 precursor on to Ago2 for miRNA maturation [53].

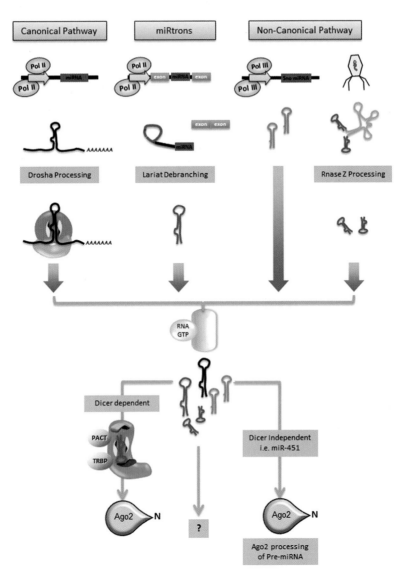

Figure 2. Alternative miRNA processing pathways in animal cells

sno-miRNAs [snoRNA (small nucleolar RNA)-derived miRNAs]

snoRNAs are an abundant class of ncRNAs mostly found in introns with diverse cellular functions. These small RNAs are transcribed by RNA Pol II and processed by exonucleolytic trimming after splicing. Typically, they guide RNA modifications and are found in nucleoli and Cajal bodies in most eukaryotic cells. The use of deep sequencing discovered that several miRNAs, such as small RNAs, were aligned to loci regions encoding the C/D and H/ACA box snoRNAs. However, so far, only one snoRNA-derived small RNA has been shown to regulate endogenous targets via a similar mechanism to miRNAs [54]. These sno-miRNAs, like the mirtrons, appear to be independent of Drosha, but may require Dicer for final maturation. It may also be possible that some sno-miRNAs are processed in the same manner as miR-451 and only require Ago2. One of the intriguing questions is why are miRNAs embedded in snoRNAs? There is speculation that sno-miRNAs may represent an ancient version of the modern day miRNA and are continuously evolving. It has also been shown that C/D and H/ACA box snoRNAs are a new family of mobile genetic elements [55]. The insertion of new snoRNA copies may be a safeguard to protect the biological activity of sno-miRNAs if and when the parental miRNA is either mutated or damaged. It is also equally plausible that many sno-miRNAs are the result of accidental processing of hairpin–like structures and have not gained function yet.

Virus-derived miRNAs

miRNAs can also be derived from pathogens such as viruses. These viral miRNAs are encoded in the viral genome and were first discovered in the herpes virus families [56]. To-date 26 mammalian viruses have been described to harbour viral miRNAs including the human EBV (Epstein–Barr virus), herpes simplex virus 1/2 [57], MDV (Marek's disease virus) [58] and KSHV (Kaposi's sarcoma-associated herpes virus) [59]. The majority of viral miRNAs from DNA viruses are processed via the canonical pathway, but the route taken by adenoviruses is somewhat different. The initial transcription is done by RNase III to generate a highly structured stem-loop precursor of 160 nt known as the VA1. This RNA is not processed by Drosha, but instead is transported by exportin-5 into the cytoplasm for Dicer processing. This processing is very inefficient with less than 1% of VA1 RNAs being processed into functional miRNAs [60]. The biological role of VA1-derived miRNAs remains to be determined, but VA1 can inhibit Dicer function and prevent the processing of cellular precursor miRNAs [60]. Another pathway used for viral miRNA biogenesis was demonstrated for the murine γ-herpes virus MHV68 [61]. This pri-miRNA structure is a combination of a tRNA-like structure linked to one or two ~60 nt pre-miRNA hairpins. This fusion RNA is transcribed by RNase III and cleaved by the enzyme RNase Z, which normally functions in the maturation of tRNA 3′-ends. This processing liberates the pre-miRNA which is then exported and further processed by Dicer in the cytoplasm. To-date the majority of studies have focused on dsDNA (double-stranded DNA) viruses and very few studies have yielded any evidence for viral miRNAs from RNA viruses. A recent study has shown that the BLV (bovine leukaemia virus), within its RNA genome, contains five miRNAs which are also transcribed by RNase III [6]. The initial transcription yields pri-miRNAs of only 55 nt, structurally similar to pre-miRNAs, and therefore bypasses Drosha processing.

Conclusion

We have accumulated a substantial amount of information regarding the processing of small RNAs, particularly miRNAs. This is due to the combination of biochemical characterizations, structural studies and high-throughput approaches such as deep sequencing protein-bound small RNAs and their intermediates. We expect to gain a greater insight into pri-miRNA processing by the crystallization of Drosha, the only key protein in the biogenesis pathway which lacks detailed structural studies. There is also increasing evidence that the regulation of the small RNA biogenesis pathway is crucial to cell-cycle regulation and responds to pathogen attack. In the future, we anticipate that more and more miRNA regulatory proteins will be discovered in specialized cell types and full animal systems. Given that there are also alternative non-canonical pathways, it is entirely plausible that different cellular or stress stimuli may govern the pathway by which specific miRNAs are processed.

Summary

- miRNA maturation is a stepwise process that starts from long structured RNAs and results in a species of 21–23 nt long single-stranded RNA which encodes their biological functions.
- To gain function the miRNAs have to be associated with an Argonaute protein forming the minimal RISC.
- The maturation of miRNAs is a highly regulated process. RNA binding and signalling proteins with function(s) in the regulation of cell cycle and immune response facilitate and/or suppress miRNA synthesis at different steps of miRNA production.
- Alternative maturation pathways exist, but these are specific to certain miRNA families.

Work of the authors is supported by the Australian Research Council and Congressionally Directed Medical Research Programs, Department of Defence. N.T. is an UTS chancellor post-doctoral fellow and G.H. is an ARC Future Fellow.

References

1. Sulston, J.E. and Brenner, S. (1974) The DNA of *Caenorhabditis elegans*. Genetics **77**, 95–104
2. Lagos-Quintana, M., Rauhut, R., Lendeckel, W. and Tuschl, T. (2001) Identification of novel genes coding for small expressed RNAs. Science **294**, 853–858
3. Lau, N.C., Lim, L.P., Weinstein, E.G. and Bartel, D.P. (2001) An abundant class of tiny RNAs with probable regulatory roles in *Caenorhabditis elegans*. Science **294**, 858–862
4. Lee, R.C. and Ambros, V. (2001) An extensive class of small RNAs in *Caenorhabditis elegans*. Science **294**, 862–864
5. Cai, X., Hagedorn, C.H. and Cullen, B.R. (2004) Human microRNAs are processed from capped, polyadenylated transcripts that can also function as mRNAs. RNA **10**, 1957–1966
6. Kincaid, R.P., Burke, J.M. and Sullivan, C.S. (2012) RNA virus microRNA that mimics a B-cell oncomiR. Proc. Natl. Acad. Sci. U.S.A. **109**, 3077–3082
7. Gregory, R.I., Yan, K.P., Amuthan, G., Chendrimada, T., Doratotaj, B., Cooch, N. and Shiekhattar, R. (2004) The Microprocessor complex mediates the genesis of microRNAs. Nature **432**, 235–240
8. Lee, Y., Ahn, C., Han, J., Choi, H., Kim, J., Yim, J., Lee, J., Provost, P., Rådmark, O., Kim, S. and Kim, V.N. (2003) The nuclear RNase III Drosha initiates microRNA processing. Nature **425**, 415–419
9. Han, J., Lee, Y., Yeom, K.H., Nam, J.W., Heo, I., Rhee, J.K., Sohn, S.Y., Cho, Y., Zhang, B.T. and Kim, V.N. (2006) Molecular basis for the recognition of primary microRNAs by the Drosha-DGCR8 complex. Cell **125**, 887–901

10. Voinnet, O. (2009) Origin, biogenesis, and activity of plant microRNAs. Cell **136**, 669–687
11. Chong, M.M., Zhang, G., Cheloufi, S., Neubert, T.A., Hannon, G.J. and Littman, D.R. (2010) Canonical and alternate functions of the microRNA biogenesis machinery. Gens Dev. **24**, 1951–1960
12. Han, J., Pedersen, J.S., Kwon, S.C., Belair, C.D., Kim, Y.K., Yeom, K.H., Yang, W.Y., Haussler, D., Blelloch, R. and Kim, V.N. (2009) Posttranscriptional crossregulation between Drosha and DGCR8. Cell **136**, 75–84
13. Fukuda, T., Yamagata, K., Fujiyama, S., Matsumoto, T., Koshida, I., Yoshimura, K., Mihara, M., Naitou, M., Endoh, H., Nakamura, T. et al. (2007) DEAD-box RNA helicase subunits of the Drosha complex are required for processing of rRNA and a subset of microRNAs. Nat. Cell Biol. **9**, 604–611
14. Davis, B.N., Hilyard, A.C., Nguyen, P.H., Lagna, G. and Hata, A. (2010) Smad proteins bind a conserved RNA sequence to promote microRNA maturation by Drosha. Mol. Cell **39**, 373–384
15. Suzuki, H.I., Yamagata, K., Sugimoto, K., Iwamoto, T., Kato, S. and Miyazono, K. (2009) Modulation of microRNA processing by p53. Nature **460**, 529–533
16. Yamagata, K., Fujiyama, S., Ito, S., Ueda, T., Murata, T., Naitou, M., Takeyama, K., Minami, Y., O'Malley, B.W. and Kato, S. (2009) Maturation of microRNA is hormonally regulated by a nuclear receptor. Mol. Cell **36**, 340–347
17. Trabucchi, M., Briata, P., Garcia-Mayoral, M., Haase, A.D., Filipowicz, W., Ramos, A., Gherzi, R. and Rosenfeld, M.G. (2009) The RNA-binding protein KSRP promotes the biogenesis of a subset of microRNAs. Nature **459**, 1010–1014
18. Michlewski, G. and Caceres, J.F. (2010) Antagonistic role of hnRNP A1 and KSRP in the regulation of let-7a biogenesis. Nat. Struct. Mol. Biol. **17**, 1011–1018
19. Guil, S. and Caceres, J.F. (2007) The multifunctional RNA-binding protein hnRNP A1 is required for processing of miR-18a. Nat. Struct. Mol. Biol. **14**, 591–596
20. Borchert, G.M., Gilmore, B.L., Spengler, R.M., Xing, Y., Lanier, W., Bhattacharya, D. and Davidson, B.L. (2009) Adenosine deamination in human transcripts generates novel microRNA binding sites. Hum. Mol. Genet. **18**, 4801–4807
21. Lund, E., Guttinger, S., Calado, A., Dahlberg, J.E. and Kutay, U. (2004) Nuclear export of microRNA precursors. Science **303**, 95–98
22. Zeng, Y. and Cullen, B.R. (2004) Structural requirements for pre-microRNA binding and nuclear export by Exportin 5. Nucleic Acids Res. **32**, 4776–4785
23. Bernstein, E., Caudy, A.A., Hammond, S.M. and Hannon, G.J. (2001) Role for a bidentate ribonuclease in the initiation step of RNA interference. Nature **409**, 363–366
24. Chendrimada, T.P., Gregory, R.I., Kumaraswamy, E., Norman, J., Cooch, N., Nishikura, K. and Shiekhattar, R. (2005) TRBP recruits the Dicer complex to Ago2 for microRNA processing and gene silencing. Nature **436**, 740–744
25. Lee, Y., Hur, I., Park, S.Y., Kim, Y.K., Suh, M.R. and Kim, V.N. (2006) The role of PACT in the RNA silencing pathway. EMBO J. **25**, 522–532
26. Liu, Q., Rand, T.A., Kalidas, S., Du, F., Kim, H.E., Smith, D.P. and Wang, X. (2003) R2D2, a bridge between the initiation and effector steps of the *Drosophila* RNAi pathway. Science **301**, 1921–1925
27. Marques, J.T., Kim, K., Wu, P.H., Alleyne, T.M., Jafari, N. and Carthew, R.W. (2010) Loqs and R2D2 act sequentially in the siRNA pathway in *Drosophila*. Nat. Struct. Mol. Biol. **17**, 24–30
28. Macrae, I.J., Zhou, K., Li, F., Repic, A., Brooks, A.N. Cande, W.Z., Adams, P.D. and Doudna, J.A. (2006) Structural basis for double-stranded RNA processing by Dicer. Science **311**, 195–198
29. Tsutsumi, A., Kawamata, T., Izumi, N., Seitz, H. and Tomari, Y. (2011) Recognition of the pre-miRNA structure by *Drosophila* Dicer-1. Nat. Struct. Mol. Biol. **18**, 1153–1158
30. Lehrbach, N.J., Armisen, J., Lightfoot, H.L., Murfitt, K.J., Bugaut, A., Balasubramanian, S. and Miska, E.A. (2009) LIN-28 and the poly(U) polymerase PUP-2 regulate let-7 microRNA processing in *Caenorhabditis elegans*. Nat. Struct. Mol. Biol. **16**, 1016–1020

31. Rybak, A., Fuchs, H., Smirnova, L., Brandt, C., Pohl, E.E., Nitsch, R. and Wulczyn, F.G. (2008) A feedback loop comprising lin-28 and let-7 controls pre-let-7 maturation during neural stem-cell commitment. Nat. Cell Biol. **10**, 987–993
32. Heo, I., Joo, C., Cho, J., Ha, M., Han, J. and Kim, V.N. (2008) Lin28 mediates the terminal uridylation of let-7 precursor microRNA. Mol. Cell **32**, 276–284
33. Newman, M.A. and Hammond, S.M. (2010) Lin-28: an early embryonic sentinel that blocks Let-7 biogenesis. Int. J. Biochem. Cell Biol. **42**, 1330–1333
34. Tokumaru, S., Suzuki, M., Yamada, H., Nagino, M. and Takahashi, T. (2008) let-7 regulates Dicer expression and constitutes a negative feedback loop. Carcinogenesis **29**, 2073–2077
35. Mueller, A.C., Sun. D. and Dutta, A. (2012) The miR-99 family regulates the DNA damage response through its target SNF2H. Oncogene **32**, 1164–1172
36. Song, J.J., Smith, S.K., Hannon, G.J. and Joshua-Tor, L. (2004) Crystal structure of Argonaute and its implications for RISC slicer activity. Science **305**, 1434–1437
37. Kwak, P.B. and Tomari, Y. (2012) The N domain of Argonaute drives duplex unwinding during RISC assembly. Nat. Struct. Mol. Biol. **19**, 145–151
38. Heyn, H., Engelmann, M., Schreek, S., Ahrens, P., Lehmann, U., Kreipe, H., Schlegelberger, B. and Beger, C. (2011) MicroRNA miR-335 is crucial for the BRCA1 regulatory cascade in breast cancer development. Int. J. Cancer **129**, 2797–2806
39. Pastrello, C., Polesel, J., Della Puppa, L., Viel, A. and Maestro, R. (2010) Association between hsa-mir-146a genotype and tumor age-of-onset in BRCA1/BRCA2-negative familial breast and ovarian cancer patients. Carcinogenesis **31**, 2124–2126
40. Khvorova, A., Reynolds, A. and Jayasena, S.D. (2003) Functional siRNAs and miRNAs exhibit strand bias. Cell **115**, 209–216
41. Schwarz, D.S., Hutvagner, G., Du, T., Xu, Z., Aronin, N. and Zamore, P.D. (2003) Asymmetry in the assembly of the RNAi enzyme complex. Cell **115**, 199–208
42. Kawamata, T. and Tomari, Y. (2010) Making RISC. Trends Biochem. Sci. **35**, 368–376
43. Tomari, Y., Matranga, C., Haley, B., Martinez, N. and Zamore, P.D. (2004) A protein sensor for siRNA asymmetry. Science **306**, 1377–1380
44. Tsuchida, T., Matsuse, H., Fukahori, S., Kawano, T., Tomari, S., Fukushima, C. and Kohno, S. (2012) Effect of respiratory syncytial virus infection on plasmacytoid dendritic cell regulation of allergic airway inflammation. Int. Arch. Allergy Immunol. **157**, 21–30
45. Betancur, J.G. and Tomari, Y. (2012) Dicer is dispensable for asymmetric RISC loading in mammals. RNA **18**, 24–30
46. Liu, J., Carmell, M.A., Rivas, F.V., Marsden, C.G., Thomson, J.M., Song, J.J., Hammond, S.M., Joshua-Tor, L. and Hannon, G.J. (2004) Argonaute2 is the catalytic engine of mammalian RNAi. Science **305**, 1437–1441
47. Johnston, M., Geoffroy, M.C., Sobala, A., Hay, R. and Hutvagner, G. (2010) HSP90 protein stabilizes unloaded argonaute complexes and microscopic P-bodies in human cells. Mol. Biol. Cell **21**, 1462–1469
48. Ruegger, S. and Grosshans, H. (2012) MicroRNA turnover: when, how, and why. Trends Biochem. Sci. **37**, 436–446
49. Okamura, K., Hagen, J.W., Duan, H., Tyler, D.M. and Lai, E.C. (2007) The mirtron pathway generates microRNA-class regulatory RNAs in *Drosophila*. Cell **130**, 89–100
50. Berezikov, E., Chung, W.J., Willis, J., Cuppen, E. and Lai, E.C. (2007) Mammalian mirtron genes. Mol. Cell **28**, 328–336
51. Havens, M.A., Reich, A.A., Duelli, D.M. and Hastings, M.L. (2012) Biogenesis of mammalian microRNAs by a non-canonical processing pathway. Nucleic Acids Res. **40**, 4626–4640
52. Cheloufi, S., Dos Santos, C.O., Chong, M.M. and Hannon, G.J. (2010) A dicer-independent miRNA biogenesis pathway that requires Ago catalysis. Nature **465**, 584–589
53. Yang, J.S., Maurin, T. and Lai, E.C. (2012) Functional parameters of Dicer-independent microRNA biogenesis. RNA **18**, 945–957

54. Ender, C., Krek, A., Friedlander, M.R., Beitzinger, M., Weinmann, L., Chen, W., Pfeffer, S., Rajewsky, N. and Meister, G. (2008) A human snoRNA with microRNA-like functions. Mol. Cell **32**, 519–528
55. Weber, M.J. (2006) Mammalian small nucleolar RNAs are mobile genetic elements. PLoS Genet. **2**, e205
56. Pfeffer, S., Zavolan, M., Grasser, F.A., Chien, M., Russo, J.J., Ju, J., John, B., Enright, A.J., Marks, D., Sander, C. and Tuschl, T. (2004) Identification of virus-encoded microRNAs. Science **304**, 734–736
57. Jurak, I., Kramer, M.F., Mellor, J.C., van Lint, A.L., Roth, F.P., Knipe, D.M. and Coen, D.M. (2010) Numerous conserved and divergent microRNAs expressed by herpes simplex viruses 1 and 2. J. Virol. **84**, 4659–4672
58. Yao, Y., Zhao, Y., Smith, L.P., Lawrie, C.H., Saunders, N.J., Watson, M. and Nair, V. (2009) Differential expression of microRNAs in Marek's disease virus-transformed T-lymphoma cell lines. J. Gen. Virol. **90**, 1551–1559
59. Cai, X., Lu, S., Zhang, Z., Gonzalez, C.M., Damania, B. and Cullen, B.R. (2005) Kaposi's sarcoma-associated herpesvirus expresses an array of viral microRNAs in latently infected cells. Proc. Natl. Acad. Sci. U.S.A. **102**, 5570–5575
60. Lu, S. and Cullen, B.R. (2004) Adenovirus VA1 noncoding RNA can inhibit small interfering RNA and microRNA biogenesis. J. Virol. **78**, 12868–12876
61. Bogerd, H.P., Karnowski. H.W., Cai, X., Shin, J., Pohlers, M. and Cullen, B.R. (2010) A mammalian herpesvirus uses noncanonical expression and processing mechanisms to generate viral microRNAs. Mol. Cell **37**, 135–142

Mechanism of miRNA-mediated repression of mRNA translation

Tamas Dalmay[1]

School of Biological Sciences, University of East Anglia, Norwich NR4 7TJ, U.K.

Abstract

MicroRNAs regulate the expression of protein-coding genes in animals and plants. They function by binding to mRNA transcripts with complementary sequences and inhibit their expression. The level of sequence complementarity between the microRNA and mRNA transcript varies between animal and plant systems. Owing to this subtle difference, it was initially believed that animal and plant microRNAs act in different ways. Recent developments revealed that, although differences still remain in the two kingdoms, the differences are smaller than first thought. It is now clear that both animal and plant microRNAs mediate both translational repression of intact mRNAs and also cause mRNA degradation.

Keywords:
degradome, microRNA, seed sequence, target recognition, translational repression.

Introduction

miRNAs (microRNAs) are negative regulators of gene expression in both plants and animals [1]. Mature miRNAs are produced by a multistep process (see Chapter 2) and are incorporated into a protein complex called the RISC (RNA-induced silencing complex). Since the mature miRNA in the RISC is single-stranded, it has the ability to anneal to another RNA molecule with a complementary sequence. miRNAs can therefore bind to specific target regions within mRNA transcripts which contain a sequence that can anneal to the miRNA. These sequences are called miRNA target sites and are usually at the 3′-UTR (untranslated region) in animals, but are in the coding region of plant mRNAs. As miRNAs bind target sites on mRNAs, they

[1]email t.dalmay@uea.ac.uk

guide the RISC to these transcripts leading to the silencing of those messages. The present chapter focuses on the mechanism of this silencing.

Target recognition

One would expect that it is relatively straightforward to identify miRNA target sites if the sequence of all of the miRNAs and mRNAs are known. However, in animals and humans it is very rare that an mRNA contains a perfect complementary target site for any miRNA [1]. The only known exception is the *HOXB8* mRNA, which is recognized by miR-196 [2]. All other human miRNAs silence mRNAs with target sites that are not perfectly complementary [1]. Identifying these target sites by a similarity search is difficult because, by allowing mismatches between miRNAs and mRNAs in a database search, hundreds or thousands of potential target sites will be identified due to the small size of the miRNAs. Furthermore, the positions of the mismatches are important and, typically, nucleotides at positions 2–8 on the miRNA are perfectly complementary to the target mRNA [3]. This region is called the seed sequence. On the basis of the number and position of mismatching nucleotides between the miRNA and its target site, three groups of target sites exist. The first group contains the 5′-dominant canonical target sites, which do not contain mismatches in the seed site and also show extensive base pairing to the rest of the miRNA. The second group is the 5′-dominant seed-only target sites, which are also perfectly complementary to the seed sequence, but only have limited base pairing with the rest of the miRNA [4]. The third group contains mismatches in the seed site, but also show extensive base pairing with the 3′ half of the miRNA. Therefore the sites in this third group are called 3′-compensatory target sites [4]. Several computational programs have been developed to predict miRNA target sites and because each of them uses a slightly different algorithm, the predicted target lists often show very little overlap with each other; e.g. TargetScan focuses on finding 5′-dominant canonical and seed-only target sites, whereas MiRanda preferentially predicts 3′-compensatory target sites [4].

It is much easier to predict miRNA targets computationally in plants because all known target sites show a very high complementarity to the entire miRNA. This does not necessarily mean that plant miRNAs cannot recognize target sites with several mismatches like in animals. However, because target sites with near perfect complementarity do exist, these have been extensively studied and little effort has been made to explore the potential targets with several mismatches.

Animal miRNAs repress translation

In order to understand how miRNAs repress translation, we first need to introduce a few features of mRNAs that are important for translation. Eukaryotic mRNAs are protected against non-specific exonuclease digestion by the cap structure at the 5′-end and by the polyA tail at the 3′-end. Efficient translation requires interaction of these two features, which leads to looping of the mRNA. The PABPC [cytoplasmic PABP (polyA-binding protein)], as the name suggests, is bound to the 3′-polyA tail, but also interacts with eIF4G (eukaryotic translation-initiation factor 4G). Because eIF4G also interacts with eIF4E, which recognizes and binds the 5′-cap structure, this sequence of interactions leads to circularization of the mRNA, which is required for efficient translation (see more on translation in [5]). Another feature of translation is the formation of polysomes. In the presence of translation inhibitors that arrest ribosomes (e.g. cycloheximide), mRNAs stay bound to ribosomes that have already initiated translation. Since more than one ribosome translates a single mRNA at once, mRNAs with various numbers of ribosomes can be purified on a sucrose gradient, these are called polysomes [6].

The first question in elucidating the mechanism of translational repression by miRNAs is to determine whether miRNAs either suppress the initiation of translation or act at the post-initiation stage. Interestingly, there is evidence for both initiation and post-initiation suppression; e.g. Petersen et al. [7] found that miRNAs which target mRNAs are present in the polysome fraction, although the protein encoded by those target mRNAs were not detectable. Since ribosomes already initiated translation of mRNAs present in the polysome fraction, it was concluded that these targets were silenced at the post-initiation stage. This model was further supported by the observation that miRNA silencing occurred in the absence of the 5′-cap structure [7]. In this experiment, an IRES (internal ribosome entry site) was used instead of the canonical 5′-cap. However, other groups obtained contradictary results showing that target mRNAs were not in the polysome fraction, but are in the ribosome-free fraction [8]. Furthermore, cap-independent translation initiated at IRESs was not silenced by miRNAs in these experiments. This suggested that miRNAs act at the initiation step and somehow inhibit ribosomes from binding to target mRNAs through a cap-dependent mechanism. This model gained further support from experiments that used cell-free extracts to study the miRNA silencing mechanism [9–13]. In these studies, miRNA silencing required an m^{7}Gppp-cap and did not affect mRNAs with an artificial Appp-cap structure, or mRNAs with an IRES [12,13].

Animal miRNAs cause target degradation

The first mRNAs identified as miRNA targets were regulated by *lin-4* in *Caenorhabditis elegans*. These mRNAs were not affected by *lin-4* at the RNA level, but only at the protein level. Therefore it was initially thought that miRNAs in animals do not degrade mRNAs, but only repress translation. However, recent transcriptome studies showed that most miRNA targets are less abundant in the presence of miRNAs and the transcript level is also increased when miRNA activity is blocked [14–17]. When siRNAs (small interfering RNAs) with perfect complementarity to their targets (or plant miRNAs with near perfect complementarity; see below) are applied to cell cultures, the target cleavage happens at a specific position between nucleotides that are annealed to the 10th and 11th nucleotides of the miRNAs [18]. However, animal and human miRNAs rarely cause cleavage at that position. Instead, a growing body of evidence indicates that miRNAs channel their targets to the cellular 5′-to-3′ mRNA decay pathway [19,20]. In this pathway mRNAs are first de-adenylated by the CAF1–CCR4–NOT deadenylase complex, followed by a DCP2 (mRNA decapping enzyme 2)-mediated de-capping (Figure 1). Loss of the 5′-cap structure allows XRN1, the major cytoplasmic 5′-to-3′ exonuclease, to degrade mRNAs (Figure 1). The evidence that miRNAs cause mRNA degradation through the cellular 5′-to-3′ mRNA decay pathway is shown by the increased level of miRNA target transcripts in the absence of components of this pathway [19,20]. Deadenylation of mRNAs by miRNAs was also detected in cell-free extracts [11,13]. The difference between results obtained

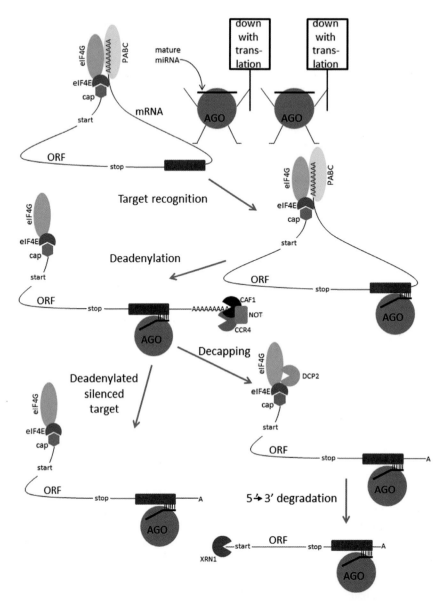

Figure 1. Translation repression and target degradation by animal miRNAs
Efficient translation of mRNAs requires a loop structure achieved by the interaction between PABPC (PABC) and eIF4G. Target sites (red rectangle) recognized by miRNAs are usually at the 3′-UTR. miRNAs are part of the AGO/Ago complex and by binding to partially complementary target sites they guide the Ago complex to target mRNAs. The Ago complex causes deadenylation of the mRNA through the CCR4–CAF1–NOT complex, which results in translational repression, owing to the lack of loop structure. Deadenylation leads to decapping followed by degradation by XRN1.

using cell cultures and cell extracts is that, *in vivo*, the deadenylated mRNAs are also decapped and then degraded but, *in vitro*, decapping and degradation does not follow deadenylation. This raised the possibility that deadenylation alone could be sufficient to silence a target mRNA [13]. Initially this was a controversial topic because deadenylation was shown to precede translational repression [11,13], but also to occur after it [21]. However, two recent studies have shown that miRNAs suppress translation initiation first, which is then followed by deadenylation and mRNA decay [22,23].

Despite these unresolved questions, it appears that animal and human miRNAs do cause target degradation, although the mechanism is different from the cleavage mechanism by plant miRNAs or siRNAs in animal cells. Since mRNA silencing involves both translational repression and mRNA degradation, it is an interesting question as to which one is more widespread. In order to answer this question one has to profile both the mRNA accumulation (to measure mRNA degradation) and protein levels (to determine translational repression). Transcriptome profiling of cells where a specific miRNA has been suppressed or overexpressed can be done using microarrays or next-generation sequencing-based RNA-seq. The proteome profiling is more challenging, but is made possible by the sophisticated method of SILAC (stable isotope labelling by amino acids in cell culture) [24]. In SILAC two populations of cells are cultured separately, one in medium containing normal amino acids and the other in medium which contains amino acids labelled with a heavy isotope. For example, the normal medium contains arginine with ^{12}C and the 'heavy medium' contains arginine with ^{13}C. Cells growing on the 'heavy medium' incorporate the ^{13}C-containing amino acids, resulting in proteins in those cells being slightly heavier than proteins in cells grown in normal medium. After protein extraction, the samples can be mixed and analysed on the same gel. This is followed by mass spectrometry, which can distinguish between the lighter and heavier form of the same protein and can provide accurate information about the ratio of the two forms. Therefore if one sample contained a higher or lower level of a specific miRNA, its effect on the proteome can be measured in SILAC experiments [24]. Two such studies have been carried out and both concluded that the extent of translation suppression by miRNAs is quite modest, usually the reduction in protein level was found to be less than 4-fold [14,15]. However, there are differences between the two studies' results concerning the dominance of one mechanism over the other. Baek et al. [15] found that very few mRNAs were repressed translationally without degradation and those targets showed weak silencing. However, Selbach et al. [14] found that at an early time point, many targets showed reduced protein levels, but unchanged mRNA levels, although most targets showed reduced mRNA and protein levels at a later time point. These differences demonstrate that more work is necessary to fully understand the extent and distribution of translational suppression and RNA degradation with respect to miRNA function. The two groups used different cell types and analysed the effects of different miRNAs so the results are not directly comparable. It is conceivable that different miRNAs could act principally through one of the pathways, and then change to the other during development or perhaps in different tissue types.

Plant miRNAs repress translation

The first two examples of translational repression of target mRNAs by plant miRNAs were the miR-172-targeted APETALA2 and miR-156/157-targeted SPL3 [25,26]. On the other hand, there were many examples of miRNA-mediated target degradation, implying that these two cases were exceptions. However, a forward genetic screen unexpectedly revealed that many plant miRNAs potentially repress translation of target mRNAs in addition to degradation [27]. The *Arabidopsis* plant that was mutated contained a transgene coding for the GFP (green fluorescent protein) with an engineered miR-171 target site in its 3′-UTR. Wild-type transgenic plants expressed the GFP protein at a low level because the endogenous miR-171 targeted the transgenic *GFP* mRNA. After mutagenesis, the seedlings were screened for increased GFP expression and two types of mutants were identified: ones that were <u>m</u>iRNA <u>b</u>iogenesis <u>d</u>eficient (*mbd*) and ones that were <u>m</u>iRNA <u>a</u>ction <u>d</u>eficient (*mad*). Two *mbd* mutants were found where miRNAs accumulated at a reduced level and these turned out to be new alleles of DCL1 (*dcl1-12*) and HEN1 (*hen1-7*). All six *mad* mutants on the other hand contained normal levels of miRNAs and could be grouped into two classes. In the first class (*mad1–4*), target mRNAs accumulated at a higher level leading to increased GFP protein levels. More interestingly, the other two *mad* mutants (*mad5* and *mad6*) contained a low level of *GFP* mRNA similar to the wild-type plants, but an increased level of GFP protein. The authors ruled out that it was a specific effect on *GFP* mRNA by showing that *GFP* mRNA without the miR-171 target site was not affected in the mutants. They also demonstrated that some endogenous target genes, which were regulated by various miRNAs, behaved similarly to the *GFP* reporter gene (where the miRNA target sites were either in the 5′-UTR, coding region or 3′-UTR). On the basis of these results it was concluded that plant miRNAs can either degrade or repress the translation of intact target mRNAs, as *mad1–4* were required for degradation and *mad5* and *mad6* were necessary for translational repression. This discovery raised several questions about plant miRNAs. Although *mad5* was identified as *KTN1*, which encodes the p60 subunit of the microtubule-severing enzyme katenin, the molecular mechanism of miRNA-mediated translational repression remains to be described. Also, *mad5* and *mad6* do not display strong phenotypic differences compared with wild-type plants. Considering that a reduced level of miRNAs in various biogenesis mutants leads to strong developmental defects, the importance and extent of the translational repression needs to be clarified. Nonetheless, it is clear that plant miRNAs can repress translation, which was further supported by a biochemical study showing that miRNAs and Ago1 can associate with polysomes [28].

Plant miRNAs cleave target mRNAs

As mentioned previously, most plant miRNAs show near perfect complementarity to their targets and cause endonucleolytic mRNA cleavage [18]. This so called 'slicing' is carried out by Ago1 and happens at the specific position between the nucleotides that are complementary to the 10th and 11th nucleotides of the miRNA (Figure 2). The cleavage generates two fragments: the 5′ fragment is protected at its 5′-end by the cap structure, but has an unprotected 3′-end, and the 3′ fragment has an exposed 5′-end and a polyA-protected 3′-end. The 5′ fragment is processed from the 3′-end by the exosome and the 3′ fragment is degraded by XRN4 [29]. Although both fragments are degraded to some extent, most 3′ fragments are more stable than their 5′ counterparts, and some of them can even be detected by Northern blot analysis [18]. Since the 3′ fragments are relatively stable and contain a monophosphate at their 5′-end, an adapter can be ligated to their 5′-end and 5′-RACE (rapid amplification of cDNA ends) analysis can be carried out to determine the exact 5′-end sequence. If the cleavage occurs exactly at the expected position (between the 10th and 11th nucleotides) of a predicted target site, this is usually accepted as an experimental proof for target validation. On the basis of this approach, a genome-wide target identification technique was developed called either degradome library sequencing [30] or PARE (parallel analysis of RNA ends) [31]. In the first step total RNA samples are enriched for

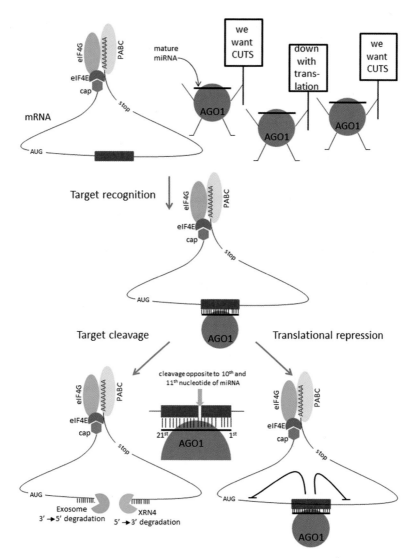

Figure 2. Translation repression and target cleavage by plant miRNAs
miRNA target sites (red rectangle) are usually in the open reading frame on plant mRNAs, although they can also be in the 5′- or 3′-UTR. Ago1-bound miRNAs anneal to near perfect complementary target sites guiding the Ago1 complex to the target mRNAs. This can result in a cleavage of mRNA at the position opposite to the 10th and 11th nucleotide of the miRNA. The cleaved mRNAs are degraded by the exosome and XRN4 (note that the degradation process by XRN4 is less efficient, therefore the 3′ cleavage fragment is usually more stable than the 5′ cleavage product). Ago1 binding can also cause translational repression without cleavage.

polyA-containing mRNAs and an adapter is ligated to the 5′-end. Then the first strand of cDNA is synthesized using an oligo(dT) primer followed by a PCR using primers that can anneal to the adapter and the oligo(dT) primer. The adapter contains a restriction site which is recognized by a restriction enzyme that cleaves away from the recognition site, exactly 21 nt downstream of the original 5′-end. After digestion with the restriction enzyme, a dsDNA (double-stranded DNA) adapter is ligated to the fragment, therefore the final product contains the 5′ adapter, the first 21 bp of the cDNA from the 3′ cleavage fragment and the 3′ adapter. Since both adaptors are compatible with the Illumina platform, the ligation products can be sequenced by a high-throughput method yielding tens of millions of sequence tags. The sequencing reads are mapped to the genome and then tested as to whether they are the 3′ halves of potential miRNA target sites and whether the 5′-end of the cleavage fragment is exactly at the expected position. The first degradome/PARE studies used the CleaveLand software [32] to analyse the sequencing results, which can identify sequence tags that map to predicted target sites of miRNAs that were given as inputs (i.e. it requires prior knowledge of small RNAs that would potentially cleave mRNAs). Using this tool normally 100–150 target mRNAs are identified for known miRNAs [31–33]. Recently a new tool was developed (PAREsnip), which does not require information about small RNAs that could potentially cleave target mRNAs [34]. PAREsnip is able to use an entire small RNA library containing tens of millions of reads to search for sequence tags mapping to cleavage sites by any small RNA. Studying *Arabidopsis* degradome libraries identified more than 4000 sites that fulfilled the strict criteria of miRNA–target site interactions and were found in two independent degradome libraries [34]. Since this number is substantially higher than the number of targets found for known miRNAs, it suggests that either there are many more unidentified miRNAs, or there are other classes of small RNAs that can cause mRNA cleavage, similar to miRNAs.

Conclusions

Over the last few years, huge advances have been made in elucidating the mechanism of miRNA-mediated mRNA targeting. Although miRNAs were initially thought to have distinct modes of function in animal and plant systems, this dogma has been revised following recent research developments. It is now clear that both mRNA degradation and translation repression occur in both kingdoms. However, the extent and occurrence of these two mechanisms has yet to be determined. There are still clear differences between plants and animals as the degradation of target protein-coding transcripts occurs through different modes-of-action. While plant miRNA complexes have slicing activity, animal miRNA complexes mediate degradation through the XRN pathway; however, plants have been shown to also use the XRN pathway to process sliced fragments in addition to the exosome. It is quite possible that both target cleavage and translational repression are used to some degree by the cell, perhaps depending on the developmental stage of the tissue in which they reside, the tissue type or the miRNA in question.

Summary
- MicroRNAs regulate the expression of protein-coding genes in animals and plants.
- Plant microRNAs cause target mRNA cleavage at a specific position, opposite to the 10th and 11th nucleotides of the microRNA.
- Plant microRNAs also cause translational suppression.
- Animal microRNAs do not cause cleavage at a specific position, but they trigger decapping and deadenylation leading to mRNA decay.
- Animal microRNAs also cause translational suppression.

References

1. Bartel, D.P. (2009) MicroRNAs: target recognition and regulatory functions. Cell **136**, 215–233
2. Yekta, S., Shih, I.H. and Bartel, D.P. (2004) MicroRNA-directed cleavage of HOXB8 mRNA. Science **304**, 594–596
3. Dalmay T. (2008) Identification of genes targeted by microRNAs. Biochem. Soc. Trans. **36**, 1194–1196
4. Sethupathy, P., Megraw, M. and Hatzigeorgiou, A.G. (2006) A guide through present computational approaches for the identification of mammalian microRNA targets. Nat. Methods **3**, 881–886
5. Jackson, R.J., Hellen, C.U. and Pestova, T.V. (2010) The mechanism of eukaryotic translation initiation and principles of its regulation. Nat. Rev. Mol. Cell Biol. **11**, 113–127
6. Ingolia, N.T., Ghaemmaghami, S., Newman, J.R. and Weissman, J.S. (2009) Genome-wide analysis *in vivo* of translation with nucleotide resolution using ribosome profiling. Science **324**, 218–223
7. Petersen, C.P., Bordeleau, M.E., Pelletier, J. and Sharp, P.A. (2006) Short RNAs repress translation after initiation in mammalian cells. Mol. Cell **21**, 533–542
8. Pillai, R.S., Bhattacharyya, S.N., Artus, C.G., Zoller, T., Cougot, N., Basyuk, E., Bertrand, E. and Filipowicz, W. (2005) Inhibition of translational initiation by Let-7 microRNA in human cells. Science **309**, 1573–1576
9. Wang, B., Love, T.M., Call, M.E., Doench, J.G. and Novina, C.D. (2006) Recapitulation of short RNA-directed translational gene silencing *in vitro*. Mol. Cell **22**, 553–560
10. Thermann, R. and Hentze, M.W. (2007) *Drosophila* miR2 induces pseudo-polysomes and inhibits translation initiation. Nature **447**, 875–878
11. Iwasaki, S., Kawamata, T. and Tomari, Y. (2009) *Drosophila* argonaute1 and argonaute2 employ distinct mechanisms for translational repression. Mol. Cell **34**, 58–67
12. Mathonnet, G., Fabian, M.R., Svitkin, Y.V., Parsyan, A., Huck, L., Murata, T., Biffo, S., Merrick, W.C., Darzynkiewicz, E., Pillai, R.S et al. (2007) MicroRNA inhibition of translation initiation *in vitro* by targeting the cap-binding complex eIF4F. Science **17**, 1764–1767
13. Wakiyama, M., Takimoto, K., Ohara, O. and Yokoyama, S. (2007) Let-7 microRNA-mediated mRNA deadenylation and translational repression in a mammalian cell-free system. Genes Dev. **21**, 1857–1862
14. Selbach, M., Schwanhäusser, B., Thierfelder, N., Fang, Z., Khanin, R. and Rajewsky, N. (2008) Widespread changes in protein synthesis induced by microRNAs. Nature **455**, 58–63
15. Baek, D., Villén, J., Shin, C., Camargo, F.D., Gygi, S.P. and Bartel, D.P. (2008) The impact of microRNAs on protein output. Nature **455**, 64–71
16. Hendrickson, D.G., Hogan, D.J., McCullough, H.L., Myers, J.W., Herschlag, D., Ferrell, J.E. and Brown, P.O. (2009) Concordant regulation of translation and mRNA abundance for hundreds of targets of a human microRNA. PLoS Biol. **7**, e1000238
17. Guo, H., Ingolia, N.T., Weissman, J.S. and Bartel, D.P. (2010) Mammalian microRNAs predominantly act to decrease target mRNA levels. Nature **466**, 835–840
18. Llave, C., Xie, Z., Kasschau, K.D. and Carrington, J.C. (2002) Cleavage of Scarecrow-like mRNA targets directed by a class of *Arabidopsis* miRNA. Science **297**, 2053–2056
19. Rehwinkel, J., Behm-Ansmant, I., Gatfield, D. and Izaurralde, E. (2005) A crucial role for GW182 and the DCP1:DCP2 decapping complex in miRNA-mediated gene silencing. RNA **11**, 1640–1647
20. Behm-Ansmant, I., Rehwinkel, J., Doerks, T., Stark, A., Bork, P. and Izaurralde, E. (2006) mRNA degradation by miRNAs and GW182 requires both CCR4:NOT deadenylase and DCP1:DCP2 decapping complexes. Genes Dev. **20**, 1885–1898
21. Fabian, M.R., Mathonnet, G., Sundermeier, T., Mathys, H., Zipprich, J.T., Svitkin, Y.V., Rivas, F., Jinek, M., Wohlschlegel, J., Doudna, J.A. et al. (2009) Mammalian miRNA RISC recruits CAF1 and PABP to affect PABP-dependent deadenylation. Mol. Cell **35**, 868–880

22. Bazzini, A.A., Lee, M.T. and Giraldez, A.J. (2012) Ribosome profiling shows that miR-430 reduces translation before causing mRNA decay in zebrafish. Science **336**, 233–237
23. Djuranovic, S., Nahvi, A. and Green, R. (2012) miRNA-mediated gene silencing by translational repression followed by mRNA deadenylation and decay. Science **336**, 237–240
24. Ong, S.E., Blagoev, B., Kratchmarova, I., Kristensen, D.B., Steen, H., Pandey, A. and Mann, M. (2002) Stable isotope labeling by amino acids in cell culture, SILAC, as a simple and accurate approach to expression proteomics. Mol. Cell. Proteomics **1**, 376–386
25. Chen, X. (2004) A microRNA as a translational repressor of APETALA2 in *Arabidopsis* flower development. Science **303**, 2022–2025
26. Gandikota, M., Birkenbihl, R.P., Höhmann, S., Cardon, G.H., Saedler, H. and Huijser, P. (2007) The miRNA156/157 recognition element in the 3′ UTR of the *Arabidopsis* SBP box gene SPL3 prevents early flowering by translational inhibition in seedlings. Plant J. **49**, 683–693
27. Brodersen, P., Sakvarelidze-Achard, L., Bruun-Rasmussen, M., Dunoyer, P., Yamamoto, Y.Y., Sieburth, L. and Voinnet, O. (2008) Widespread translational inhibition by plant miRNAs and siRNAs. Science **320**, 1185–1190
28. Lanet, E., Delannoy, E., Sormani, R., Floris, M., Brodersen, P., Crété, P., Voinnet, O., and Robaglia, C. (2009) Biochemical evidence for translational repression by *Arabidopsis* microRNAs. Plant Cell **21**, 1762–1768
29. Souret, F.F., Kastenmayer, J.P. and Green, P.J. (2004) AtXRN4 degrades mRNA in *Arabidopsis* and its substrates include selected miRNA targets. Mol. Cell **15**, 173–183
30. Addo-Quaye, C., Eshoo, T.W., Bartel, D.P. and Axtell, M.J. (2008) Endogenous siRNA and miRNA targets identified by sequencing of the *Arabidopsis* degradome. Curr. Biol. **18**, 758–762
31. German, M.A., Pillay, M., Jeong, D.H., Hetawal, A., Luo, S., Janardhanan, P., Kannan, V., Rymarquis, L.A., Nobuta, K., German, R. et al. (2008) Global identification of microRNA-target RNA pairs by parallel analysis of RNA ends. Nat. Biotechnol. **26**, 941–946
32. Addo-Quaye, C., Miller, W. and Axtell, M.J. (2009) CleaveLand: a pipeline for using degradome data to find cleaved small RNA targets. Bioinformatics **25**, 130–131
33. Pantaleo, V., Szittya, G., Moxon, S., Miozzi, L., Moulton, V., Dalmay, T. and Burgyan, J. (2010) Identification of grapevine microRNAs and their targets using high-throughput sequencing and degradome analysis. Plant J. **62**, 960–976
34. Folkes, L., Moxon, S., Woolfenden, H.C., Stocks, M.B., Szittya, G., Dalmay, T. and Moulton, V. (2012) PAREsnip: a tool for rapid genome-wide discovery of small RNA/target interactions evidenced through degradome sequencing. Nucleic Acids Res. **40**, e103

© The Authors Journal compilation © 2013 Biochemical Society
Essays Biochem. (2013) 54, 39–52: doi: 10.1042/BSE0540039

Piwi-interacting RNAs: biological functions and biogenesis

Kaoru Sato*[1] and Mikiko C. Siomi*[†1,2]

*Department of Biophysics and Biochemistry, Graduate School of Science, The University of Tokyo, Tokyo, Japan
†Core Research for Evolutional Science and Technology, Japan Science and Technology Agency, Saitama, Japan

Abstract

The integrity of the germline genome must be maintained to achieve successive generations of a species, because germline cells are the only source for transmitting genetic information to the next generation. Accordingly, the germline has acquired a system dedicated to protecting the genome from 'injuries' caused by harmful selfish nucleic acid elements, such as TEs (transposable elements). Accumulating evidence shows that a germline-specific subclass of small non-coding RNAs, piRNAs (piwi-interacting RNAs), are necessary for silencing TEs to protect the genome in germline cells. To silence TEs post-transcriptionally and/or transcriptionally, mature piRNAs are loaded on to germline-specific Argonaute proteins, or PIWI proteins, to form the piRISC (piRNA-induced silencing complex). The present chapter will highlight insights into the molecular mechanisms underlying piRISC-mediated silencing and piRNA biogenesis, and discuss a possible link with tumorigenesis, particularly in *Drosophila*.

Keywords:

Drosophila, germline, piwi-interacting RNA, transposable element.

[1]Correspondence may be addressed to either author (email kaoru@biochem.s.u-tokyo.ac.jp or siomim@biochem.s.u-tokyo.ac.jp).

Introduction

RNA silencing is achieved by gene silencing pathways that are mediated by small ncRNAs (non-coding RNAs) of 20–30 nt in length. The effector complex in RNA silencing is the RISC (RNA-induced silencing complex), which consists of a member of the Argonaute family of proteins and a small RNA that guides the RISC to its targets to be silenced [1–6]. Upon RISC recognition of targets, Argonaute protein inhibits their expression by either cleaving them with its Slicer endonuclease activity, or by inducing translational inhibition, RNA destabilization and chromatin remodelling through DNA methylation and/or histone modification.

piRNAs (piwi-interacting RNAs) are a subset of small RNAs that trigger RNA silencing in the gonads. piRNAs are typically 23–30 nt long and associate specifically with germline-specific Argonaute proteins of the PIWI subfamily [7]. piRNAs were discovered during an investigation into how the tandemly repeated *Ste* (*Stellate*) gene is silenced in the male germline of *Drosophila melanogaster* [8]. Without *Ste* silencing, spermatogenesis does not proceed properly and males become infertile. *Ste* silencing was linked to a Y-chromosomal locus, *Su(Ste)* (*Suppressor of Stellate*), which consists of tandem repeats showing strong similarity to *Ste* at the nucleotide sequence level [6,8]. The functional importance of small RNAs arising from *Su(Ste)* was argued in early 2000 [8] when the mechanism underlying *Ste* silencing was still unclear. Later, a small RNA profiling study on *Drosophila* testes and early embryos revealed endogenous 23–30 nt small RNAs derived from repetitive intergenic elements scattered across the genome, including from *Su(Ste)* and TEs (transposable elements) [9]. These small RNAs were originally termed rasiRNAs [repeat-associated siRNAs (small interfering RNAs)], but they are currently known as piRNAs because they associate with PIWI proteins to function in invertebrate and vertebrate RNA silencing [10–15].

In the present chapter, we summarize our current understanding of piRNA functions and biogenesis, mainly focusing on studies using *D. melanogaster*. We also review recent progress in our understanding of the biological involvement of the piRNA pathway in non–gonadal cells, such as in brain tumour development.

Targets of piRNA-mediated silencing

TE silencing by the piRISC (piRNA-induced silencing complex)

In *Drosophila*, piRNAs associate with three PIWI proteins, Piwi, Aub and AGO3, to form the piRISC and guide the silencing of RNAs that accommodate complementary sequences. The main targets of piRNAs are TEs and this is highly conserved across animal species. TEs are genomic parasites that move from one chromosomal location to another by either a cut-and-paste (transposition) or copy-and-paste (insertion) mode of action [16]. In this way, they potentially modify and disrupt the functions of other genes and often threaten the integrity of the host genome [6,16]. Studies using *Drosophila* oogenesis have provided critical insights into the molecular mechanisms underlying piRNA biogenesis and piRNA-mediated TE silencing.

Mutations in *piwi*, *aub* or *ago3* lead to TE derepression in the germline, indicating the non-redundancy of these genes [6,17–23]. Indeed, all *PIWI* genes have crucial roles in gonadal development: both *piwi* and *aub* are required for male and female fertility [24,25], whereas *ago3* is required for female fertility, but only partially required for male fertility [18]. The non-redundancy of PIWI proteins may also be explained by the fact that, unlike germline-specific Aub and Ago3, *piwi* is expressed in both germline and somatic cells [19–22,26].

PIWI proteins exhibit Slicer activity *in vitro*; thus piRISCs probably silence their targets post-transcriptionally, by cleaving them, as does Ago2 associated with siRNAs. Aub and AGO3 are involved in Slicer-mediated silencing in the cytoplasm (see below for more detail). Unlike Aub and Ago3, Piwi is localized in the nucleus. Also, nuclear localization, but not Slicer activity of Piwi, is required for silencing TEs [26,27]. In mice, DNA methylation influences chromatin structures and is strongly related to piRNA-mediated RNA silencing in testes [28], leading to speculation that Piwi in flies would also induce DNA methylation or covalent histone modifications in the nucleus. Yet, DNA methylation activity has not been conclusively detected in *Drosophila* gonads.

Protein-coding gene silencing by the piRISC

Some piRNAs are proposed to target protein-coding genes, such as *FasIII* (*Fasciclin III*) and *nos* (*nanos*) [20,27,29]. The mRNA level of *FasIII* is up-regulated in *tj* (*traffic jam*) mutants, which show a failure of intermingling of germline cells with surrounding somatic cells in the larval ovary, leading to infertility [30]. The *tj* gene encodes a large Maf transcription factor. Interestingly, piRNAs are produced from the 3′-UTR (untranslated region) of *tj* mRNAs, and these form a piRISC with Piwi. In *piwi* mutants, *FasIII* is up-regulated, implying that *tj*-piRNAs silence *FasIII* in collaboration with Piwi [27]. piRNAs may also induce degradation of maternally deposited mRNAs. *nos* encodes a posterior morphogen important for *Drosophila* germline development. *nos* mRNA deadenylation and decay are likely to be regulated by piRNAs derived from *roo* and *412* transposons, which show complementarity to the *nos* 3′-UTR. Smaug, an RNA-binding protein that provides translational repression of unlocalized *nos* mRNA, may assist by recruiting the CCR4 deadenylation complex to the target during maternal-zygotic transition in the embryo [29].

piRNA biogenesis

In animals, endogenous siRNAs also silence TEs [1,3]; however, in contrast with non-gonadal somatic cells where endogenous siRNAs are the main trigger of TE silencing, piRNAs in the germline function at the forefront of the defence against transposons. The current model of piRNA biogenesis involves two spatially and mechanistically distinct pathways: the primary processing pathway and a secondary amplification pathway (ping-pong amplification loop) (Figure 1). In the germline, piRNA biogenesis involves both pathways, whereas piRNAs in the gonadal somas are generated solely via the primary processing pathway.

The primary processing pathway

Primary piRNAs are produced from the piRNA clusters that act as sources of piRNAs and that are often located in pericentromeric or subtelomeric regions on the genome [6,26]. Each cluster spans several to more than 200 kb and contains multiple transposon fragments [6,26]. Most of the piRNA clusters produce piRNAs from both genomic strands, suggesting bidirectional transcription [6,26]. Other clusters, such as *flam* (*flamenco*), produce piRNAs almost exclusively from one genomic strand [26]. Investigation of piRNA biogenesis in *Drosophila* ovarian

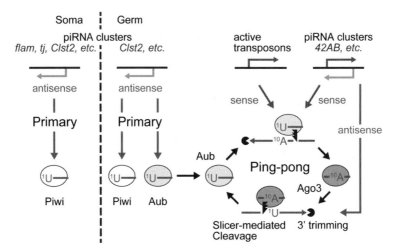

Figure 1. Two piRNA biogenesis pathways
In the *Drosophila melanogaster* primary piRNA processing pathway, antisense transcripts transcribed from piRNA clusters and/or transposons are processed to piRNAs by unknown mechanisms and are loaded on to Aub or Piwi. piRNAs derived from the *flam* locus are exclusively loaded on to Piwi because *flam* is active only in ovarian somas where only Piwi is expressed. piRISCs produced through this mechanism act as a 'trigger' for the amplification loop. The amplification loop (also known as the ping–pong cycle) most probably involves the Slicer activity of Aub and Ago3, but not that of Piwi. Aub associated with antisense piRNA cleaves complementary piRNA precursors (sense). This determines the 5′-ends of piRNAs that are loaded on to Ago3. Ago3 associated with sense piRNA cleaves complementary piRNA precursors, generating the 5′-end of antisense piRNAs, which are subsequently loaded on to Aub. The 3′-ends of piRNAs are trimmed by an unknown nuclease (or nucleases), which is followed by 2′-O-methylation mediated by HEN1/Pimet. piRNAs that induce the amplification loop may also be maternally deposited.

somas, in which only Piwi is expressed, revealed the molecular mechanism of the primary processing pathway. Representatives of primary piRNAs are *flam*-piRNAs, which are predominantly expressed in ovarian somatic cells (Figure 1) [19]. *flam*-piRNAs are mostly antisense to active transposons, and thus act as *trans*-silencers of TEs. Studies have identified two putative RNA helicases, Armi and FS(1)Yb [Female Sterile (1) Yb; also known as Yb], and a nuclease, Zuc (Zucchini), as primary piRNA factors (Figure 2). A lack of Armi or Yb eliminates Piwi-associated primary piRNAs from somatic follicle cells of ovaries and cell cultures, the ovarian somatic sheet and ovarian somatic cells [31–34]. Armi and Yb are components of Yb bodies, non-membranous high-density structures adjacent to mitochondria [31,34]. Depletion of Yb causes the disappearance of Yb bodies (Figure 2), which interferes with primary piRNA production and piRISC formation, causing mislocalization of Piwi in the cytoplasm, leading to derepression of TEs [31–34]. Therefore Yb bodies can be considered as the cytoplasmic centre for piRNA production, piRISC formation and inspection. Zuc was originally identified as a gene required for axis determination during oogenesis, and was also found to be required for piRNA biogenesis in ovaries and testes. Zuc encodes an endoribonuclease which localizes on the surface of mitochondria (Figure 2). A lack of Zuc eliminates Piwi-associating primary piRNAs, and leads to the accumulation of piRNA precursor- or intermediate-like molecules, suggesting that piRNA intermediates derived from uni-strand clusters such as *flam* are cleaved by Zuc to produce piRNAs that bind to Piwi [27,34,35]. In germline cells, primary piRNAs are

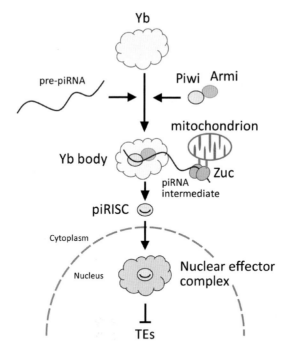

Figure 2. Primary piRNA biogenesis in follicle cells of the ovary
In follicle cells, Piwi, but not Aub or Ago3, is expressed. Armi associates with Piwi and localizes it to Yb bodies. Yb is the main component of Yb bodies. The piRNA intermediates (intermediate), partially processed from the primary piRNA precursors (pre-piRNA), are loaded on to the complex comprising Armi, Piwi and Yb at Yb bodies and are processed into mature piRNAs. Zuc localizes on the surface of mitochondria and is also required for primary piRNA processing. Without piRNA loading, Piwi is not localized to the nucleus. The piRISC complex may be associated with other nuclear proteins to form a larger functional complex (nuclear effector complex) to repress TEs in the nucleus.

required to initiate the amplification loop, although the protein factors required for the primary pathway in the germline remain unknown.

The ping-pong amplification loop

As in ovarian somas, in the germline, the primary processing pathway provides an initial pool of piRNAs. Primary piRNAs arising from bidirectional piRNA clusters, such as the *42AB* cluster, which contains transposon fragments in both sense and antisense orientations, act as sources for the ping-pong amplification loop. The ping-pong amplification loop further shapes the piRNA population by amplifying sequences that target active transposons. The ping-pong amplification loop is conserved in many animal species [6,14,15].

The ping-pong amplification loop requires pre-existing primary piRNAs, which are mainly antisense against active transposon transcripts (Figure 1). These piRNAs preferentially associate with two PIWI subfamily proteins, Piwi and Aub [6,19,26]. In contrast, sense-strand piRNAs preferentially associate with Ago3 [6,22,26]. Antisense piRNAs bound to Piwi or Aub show a strong bias for 1U-containing piRNAs, whereas sense-strand piRNAs bound to Ago3 tend to have an adenine at position 10 [6,22,26]. piRNAs that are associated with Aub or Ago3 often overlap at their 5′-ends by 10 nt. These findings led to the ping-pong amplification loop

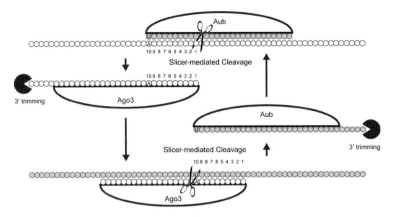

Figure 3. The ping-pong amplification loop
Aub associates predominantly with 1U-piRNAs arising from antisense transcripts of transposons. The Aub-piRNA complex guides Slicer-dependent cleavage of sense TE transcripts, yielding sense piRNA precursors with 10A. How the 3′-end of piRNAs is formed remains unknown. Reciprocally, Ago3-associated piRNAs guide cleavage of antisense TE transcripts, yielding antisense piRNA precursors with 1U.

model for piRNA biogenesis, where antisense piRNAs with Aub direct the cleavage of sense-strand transposon transcripts, generating sense piRNAs for Ago3 (Figure 3). The resulting Ago3–piRNA complex then directs cleavage of antisense piRNA precursors, generating antisense piRNAs for Aub. The 5′-ends of amplified secondary piRNAs are determined by Aub and Ago3 Slicer. The factors necessary for 3′ trimming in this cycle remain undetermined. The ping-pong amplification loop thus obviates the need for an RdRP (RNA-dependent RNA polymerase), which is needed to amplify siRNA triggers in plants, nematodes and yeast [1]. In fact, genes encoding RdRP are not found in the *Drosophila* genome.

piRNA pathway factors in *Drosophila*
Factors for piRNA cluster transcription

The mechanism underlying transcription of piRNA clusters is still an important question to be answered in the piRNA field, although several proteins have been shown to be involved in this process. *Rhi* (*Rhino*), which encodes an HP1 (heterochromatin protein 1) homologue, associates with the *42AB* cluster and this association is required for its transcription and for secondary piRNA accumulation specifically in germline cells (Table 1) [36]. *cuff* (*cutoff*), a gene related to the yeast transcription termination factor *Rai1*, physically interacts with Rhi and accumulates at centromeric/pericentromeric regions in the nucleus of germline cells and strongly co-localizes with the major heterochromatic domains [37]. Cuff is required for *42AB* cluster transcription, but some regions in the *42AB* cluster are transcribed by a Cuff-independent mechanism [37], suggesting that the dual-strand cluster might not produce single full-length transcripts spanning the entire locus, as opposed to the *flam* locus [26]; rather, the *42AB* locus may have multiple internal promoters.

A histone methyltransferase, dSETDB1, responsible for H3K9me3 (histone H3 Lys^9 trimethylation), is required for both bi- and uni-directional piRNA cluster transcription in both germline and somatic cells of the gonads (Table 1) [39], indicating that H3K9me3 is a unifying

Table 1. Factors involved in piRNA biogenesis in *Drosophila*

B-box, zinc finger B-box domain; CHROMO, CHRromatin Organization MOdifier domain; CHROMO shadow, a variant of CHROMO; DEAD, DEAD-box helicase; DEXDc, DEAD-like helicase domain; EGFP, enhanced green fluorescent protein; HA, haemagluttinin; HA2, helicase-associated domain; HELICc, helicase-superfamily C-terminal domain; KH, K homology domain; MBD, Methyl-CpG binding domain; MID, middle; MYND, zinc-finger myeloid-nervy-DEAF-1 domain; N.D., not determined; PAZ, PIWI/Argonaute/Zwille; PIWI, P-element-induced whimpy testes; PreSET, N-terminal domain to SET; RING, Really Interesting New Gene finger domain; SAM, S-adenosylmethionine; SET, Su(var)3-9/Enhancer-of-zeste/Trithorax domain; SNase, Staphylococcal nuclease; UBA, ubiquitin associated domain; -, undetected.

Gene name	Symbol	CG number	TUDOR domains	Other domains	Category	Mammalian homologue	Cellular localization Germ*	Germline- soma†	Interaction with PIWIs	Other binding partners‡	Reference(s)
piwi	piwi	CG6122		PAZ, MID, PIWI	PIWI subfamily proteins	PIWIL1–4	Nucleus	Nucleus	None	Yb, Armi, Zuc	[26,57]
aurbergine	aub	CG6137		PAZ, MID, PIWI	PIWI subfamily proteins	PIWIL1–4	Nuage	–	Ago3	Tud, Vas, Spn-E, Tej	[26,57]
argonaute 3	ago3	CG40300		PAZ, MID, PIWI	PIWI subfamily proteins	PIWIL1–4	Nuage	–	Aub	Tud	[5,18,26]
dPRMT5	dPRMT5	CG3730		SAM-methyltransferase	Arginine methylation	PRMT5	N.D.	N.D.	N.D.	N.D.	[44]
valois	vls	CG10728		4 × WD	Arginine methylation	MEP50	Nuage	–	N.D.	dPRMT5, Tud	[58]
partner of piwis	papi	CG7082	1	2 × KH	TDRDs	TDRD2	Nuage	–	Piwi, Ago3	Tral, Me31B, TER94	[41]

(Continued)

Table 1. Factors involved in piRNA biogenesis in Drosophila (Continued)

Gene name	Symbol	CG number	TUDOR domains	Other domains	Category	Mammalian homologue	Cellular localization Germ*	Germline-soma†	Interaction with PIWIs	Other binding partners‡	Reference(s)
qin (kumo)	qin (kumo)	CG14303, CG14306	5	RING, B-box	TDRDs, piRNA cluster transcription	TDRD4	Nuage, nucleus	–	Piwi, Aub	Vas, Spn-E, HP1	[40,42]
tejas	tej	CG8589	1	Lotus (Tejas)	TDRDs	TDRD5	Nuage	–	Aub	Vas, Spn-E	[43]
tudor	tud	CG9450	11		TDRDs	TDRD6	Nuage	Cytoplasm	Aub, Ago3	N.D.	[44,57]
spindle-E	spn-E	CG3158	1	DEXDc, HELICc, HA2	TDRDs	TDRD9	Nuage		Aub	Vas, Tej	[38]
FS(1)Yb	Yb	CG2706	1	Helicase-like, DEAD, Helicase, ZF	TDRDs	TDRD12	–	Yb body	Piwi	Armi, Vret	[31–34]
Brother of Yb	BoYb	CG11133	1	DEAD, Helicase, ZF	TDRDs	TDRD12	Nuage	–	N.D.	N.D.	[45]
Sister of Yb	SoYb	CG31755	2	DEAD, Helicase, ZF	TDRDs	TDRD12	Cytoplasm	Yb body	N.D.	N.D.	[45]
krimper	krimp	CG15707	1		TDRDs	None	Nuage	N.D.	N.D.	N.D.	[38,57]
vreteno	vret	CG4771	2		TDRDs	None	Nuage	Yb body	Piwi, Aub, Ago3	Yb, Armi	[46]
dSETDB1	dSETDB1	CG12196	2	MBD, PreSET, 2×SET	TDRDs, piRNA cluster transcription	SETDB1	Nucleus	Nucleus	N.D.	N.D.	[39]

Gene	Symbol	CG number	Domain/motif	Function	Homolog	Localization*	Localization†	Interacting partner*	Interacting partner†	Ref.
rhino	rhi	CG10683	CHROMO, CHROMO shadow	piRNA cluster transcription	HP1d	Nucleus	-	N.D.	Cuff	[36]
cutoff	cuff	CG13190	RAI1-like	piRNA cluster transcription	DOM3Z	Nuage (triple HA-tagged), nucleus (EGFP-tagged)	-	N.D.	Rhi	[37]
vasa	vas	CG43081	DEADc, HELICc	?	VASA	Nuage	-	Aub	Tej, Spn-E	[38]
armitage	armi	CG11513	Helicase	?	MOV10	Cytoplasm	Yb body	Piwi	Yb, Vret	[22–31]
zucchini	zuc	CG12314	Phospholipase-D/nuclease family	Endoribonuclease	mitoPLD	Nuage	N.D.	Aub	N.D.	[35,50]
maelstrom	mael	CG11254	HMG-box	Nuclease?	Maelstrom	Nuage, nucleus, spindle MT	Cytoplasm, nucleus, spindle MT	none	MTOC proteins	[49]
squash	squ	CG4711	Similar to RNase HII	Nuclease?	None	Nuage	N.D.	Aub	N.D.	[50]
heat shock protein 83	hsp83	CG1242	Hsp90, ATPase	Chaperon	Hsp90	Any subcellular fraction	Any subcellular fraction	Piwi	HOP	[52,53]

*Nurse cells
†Follicle cells
‡Only proteins relating to the piRNA or other small ncRNA biogenesis pathways.

mark dictating piRNA cluster transcription in both germline and somatic cells. Rhi recognizes H3K9me3 at bidirectional germline clusters, but there must be an alternative Rhi-independent mechanism directing somatic and unidirectional germline cluster transcription. These observations strongly indicate that repressive marks deposited by dSETDB1 are required for transcription from all major piRNA clusters, although how these marks are targeted to piRNA clusters and how repressive marks act in the transcription of piRNA clusters remain unknown.

Other piRNA pathway factors

Many TDRD (Tud domain-containing) proteins, including Tudor, Spindle-E, Krimper, Tejas, Vreteno, Qin/Kumo and PAPI (Partner of PIWIs) are required for piRNA-mediated TE silencing in germline cells, most likely for the amplification loop (Table 1) [40–46]. Both genetic and biochemical studies have indicated that TDRD proteins associate with PIWI proteins through their sDMAs (symmetric dimethylarginine residues) and that the responsible factor for the sDMA modification of PIWI proteins is DART5/PRMT5 (protein arginine N-methyltransferase 5) [6,47]. Tudor acts as a 'scaffold' facilitating the piRNA amplification pathway by recruiting sDMA-modified Aub and Ago3 and their targets in germline cells [44]. Functions of other TDRD proteins remain largely undetermined.

A number of proteins other than TDRD proteins have been genetically implicated in piRNA biogenesis. Mael (Maelstrom), a HMG (high mobility group)-box protein, is a piRNA factor (Table 1) [38,48]. A recent study showed that Mael co-ordinates microtubule organization via interacting with protein components of the MTOC (microtubule-organizing centre) [49]; however, it remains unknown at the molecular level how microtubule organization is physically connected with piRNA biogenesis. Mutations in a helicase, Vasa, and in a putative nuclease, Squash, also have an impact on piRNA populations in ovaries (Table 1) [6,50]. These piRNA factors are mostly localized at the nuage [6,38,50]. Functional alterations of Hsp (heat-shock protein) 90, which have been previously implicated in canalization [51], affect the piRNA biogenesis pathway, leading to transposon activation and the induction of morphological mutants (Table 1) [52]. Hsp90 forms a complex with Piwi and an Hsp70/Hsp90 organizing protein homologue, HOP (Table 1) [53]. Post-translational regulation of Piwi by Hsp90 and HOP may allow Piwi to suppress the generation of new genotypes by transposon-mediated 'canonical' mutagenesis [51–53]. The precise molecular functions of many piRNA factors await further investigation.

The possible role(s) of piRNAs in cancer

A growing number of studies have found that PIWI proteins in humans and mice, specifically, HIWI, PIWIL2 and PIWIL2-like proteins, are expressed in various types of tumour cells [54,55]. In addition, piRNAs were also detected in these cells [55]. These results indicate that cancer development may be linked to the piRNA pathway. A study demonstrated that ectopic expression of piRNA pathway genes contributes to the growth and development of malignant brain tumours in *Drosophila* [56]. Inactivation of the germline genes *vasa*, *piwi* or *aub* suppressed malignant tumour growth, demonstrating that germline traits are necessary for tumour growth, at least in *Drosophila* [56]. Although the potential role of piRNAs in cancer has just emerged and remains to be investigated, these data highlight the importance of understanding the exact role of the piRNA pathway during tumorigenesis and suggest new possibilities for tumour therapy.

© The Authors Journal compilation © 2013 Biochemical Society

Conclusions

The system that piRNAs utilize to protect the genome in the gonads against harmful TEs is elaborate. Recent detailed investigations of the piRNA machinery have revealed that piRNAs are mostly produced through a very dedicated, rather complex, system that requires many factors. New factors are still emerging, and a full understanding of piRNA biogenesis is still a long way off. Thus we still struggle to see fully how piRNAs, especially those in the nucleus, silence TEs at the molecular level. The results of recent studies suggest that piRNAs might also be involved in regulating stem cell and cancer development outside of the gonads. Further studies with more sophisticated techniques than those used to date will be required to achieve our goal and to potentially expand our use of piRNAs as tools in research and therapy.

Summary
- piRNAs (piwi-interacting RNAs) are a germline-specific class of small non-coding RNAs.
- piRNAs protect the integrity of germline cell genomes from harmful transposons.
- There are two plausible models for piRNA biogenesis, the primary piRNA biogenesis pathway and the ping-pong amplification loop.
- Many proteins, such as Tudor domain-containing proteins and heterochromatin-related proteins, are involved in the piRNA biogenesis pathway.
- piRNAs and PIWI subfamily proteins are expressed in somatic stem cells and several cancer cells, and might regulate stem cell and cancer development.

We apologize to colleagues whose relevant primary publications were not cited because of space constraints. We thank all the members of the Siomi laboratory at Keio University School of Medicine, especially K. Saito and H. Siomi, for useful comments and critical reading of the chapter before submission. Work in our laboratory is supported by grants from the Ministry of Education, Culture, Sports, Science and Technology (MEXT, Japan) and CREST (JST, Japan).

References
1. Ghildiyal, M. and Zamore, P.D. (2009) Small silencing RNAs: an expanding universe. Nat. Rev. Genet. **10**, 94–108
2. Siomi, H. and Siomi, M.C. (2009) On the road to reading the RNA-interference code. Nature **457**, 396–404
3. Kim, N.V., Han, J. and Siomi M.C. (2009) Biogenesis of small RNAs in animals. Nat. Rev. Mol. Cell Biol. **10**, 126–139
4. Malone, C.D. and Hannon, G.J. (2009) Small RNAs as guardians of the genome. Cell **136**, 656–668
5. Thomson, T. and Lin, H. (2009) The biogenesis and function of PIWI proteins and piRNAs: progress and prospect. Annu. Rev. Cell Dev. Biol. **25**, 355–376

6. Siomi, M.C., Sato, K., Pezic, C. and Aravin, A.A. (2011) PIWI-interacting small RNAs: the vanguard of genome defence. Nat. Rev. Mol. Cell Biol. **12**, 246–258
7. Vagin, V.V., Sigova, A., Li, C., Seitz, H., Gvozdev, V. and Zamore, P.D. (2006) A distinct small RNA pathway silences selfish genetic elements in the germline. Science **313**, 320–324
8. Aravin, A.A., Naumova, N.M., Tulin, A.V., Vagin, V.V., Rozovsky, Y.M. and Gvozdev, V.A. (2001) Double-stranded RNA-mediated silencing of genomic tandem repeats and transposable elements in the *D. melanogaster* germline. Curr. Biol. **11**, 1017–1027
9. Aravin, A.A., Lagos-Quintana, M., Yalcin, A., Zavolan, M., Marks, D., Snyder, B., Gaasterland, T., Meyer, J. and Tuschl, T. (2003) The small RNA profile during *Drosophila melanogaster* development. Dev. Cell **5**, 337–350
10. Carmell, M.A., Girard, A., van de Kant, H.J., Bourc'his, D., Bestor, T.H., de Rooij, D.G. and Hannon, G.J. (2007) MIWI2 is essential for spermatogenesis and repression of transposons in the mouse male germline. Dev. Cell **12**, 503–514
11. Aravin, A.A., Sachidanandam, R., Girard, A., Fejes-Toth, K. and Hannon, G.J. (2007) Developmentally regulated piRNA clusters implicate MILI in transposon control. Science **316**, 744–747
12. Kuramochi-Miyagawa, S., Watanabe, T., Gotoh, K., Totoki, Y., Toyoda, A., Ikawa, M., Asada, N., Kojima, K., Yamaguchi, Y., Ijiri, T.W. et al. (2008) DNA methylation of retrotransposon genes is regulated by Piwi family members MILI and MIWI2 in murine fetal testes. Genes Dev. **22**, 908–917
13. Lau, N.C., Ohsumi, T., Borowsky, M., Kingston, R.E. and Blower, M.D. (2009) Systematic and single cell analysis of *Xenopus* Piwi-interacting RNAs and Xiwi. EMBO J. **28**, 2945–2958
14. Lau, N.C., Seto, A.G., Kim, J., Kuramochi-Miyagawa, S., Nakano, T., Bartel, D.P. and Kingston, R.E. (2006) Characterization of the piRNA complex from rat testes. Science **313**, 363–367
15. Robine, N., Lau, N.C., Balla, S., Jin, Z., Okamura, K., Kuramochi-Miyagawa, S., Blower, M.D. and Lai, E.C. (2009) A broadly conserved pathway generates 3′UTR-directed primary piRNAs. Curr. Biol. **19**, 2066–2076
16. Kazazian, Jr, H.H. (2004) Mobile elements: drivers of genome evolution. Science **303**, 1626–1632
17. Saito, K. and Siomi, M.C. (2010) Small RNA-mediated quiescence of transposable elements in animals. Dev. Cell **19**, 687–697
18. Li, C., Vagin, V.V., Lee, S., Xu, J., Ma, S., Xi, H., Seitz, H., Horwich, M.D., Syrzycka, M., Honda, B.M. et al. (2009) Collapse of germline piRNAs in the absence of Argonaute3 reveals somatic piRNAs in flies. Cell **137**, 509–521
19. Malone, C.D., Brennecke, J., Dus, M., Stark, A., McCombie, W.R., Sachidanandam, R. and Hannon, G.J. (2009) Specialized piRNA pathways act in germline and somatic tissues of the *Drosophila* ovary. Cell **137**, 522–535
20. Nishida, K.M., Saito, K., Mori, T., Kawamura, Y., Nagami-Okada, T., Inagaki, S., Siomi, H. and Siomi, M.C. (2007) Gene silencing mechanisms mediated by Aubergine piRNA complexes in *Drosophila* male gonad. RNA **13**, 1911–1922
21. Saito, K., Nishida, K.M., Mori, T., Kawamura, Y., Miyoshi, K., Nagami, T., Siomi, H. and Siomi, M.C. (2006) Specific association of Piwi with rasiRNAs derived from retrotransposon and heterochromatic regions in the *Drosophila* genome. Genes Dev. **20**, 2214–2222
22. Gunawardane, L.S., Saito, K., Nishida, K.M., Miyoshi, K., Kawamura, Y., Nagami, T., Siomi, H. and Siomi, M.C. (2007) A slicer-mediated mechanism for repeat-associated siRNA 5′ end formation in *Drosophila*. Science **315**, 1587–1590
23. Nagao, A., Mituyama, T., Huang, H., Chen, D., Siomi, M.C. and Siomi, H. (2010) Biogenesis pathways of piRNAs loaded onto AGO3 in the *Drosophila* testis. RNA **16**, 2503–2515
24. Cox, D.N., Chao, A., Baker, J., Chang, L., Qiao, D. and Lin, H. (1998) A novel class of evolutionarily conserved genes defined by piwi are essential for stem cell self-renewal. Genes Dev. **12**, 3715–3727
25. Schmidt, A., Palumbo, G., Bozzetti, M.P., Tritto, P., Pimpinelli, S. and Schäfer, U. (1999) Genetic and molecular characterization of sting, a gene involved in crystal formation and meiotic drive in the male germ line of *Drosophila melanogaster*. Genetics **151**, 749–760

26. Brennecke, J., Aravin, A.A., Stark, A., Dus, M., Kellis, M., Sachidanandam, R. and Hannon, G.J. (2007) Discrete small RNA-generating loci as master regulators of transposon activity in Drosophila. Cell **128**, 1089–1103
27. Saito, K., Inagaki, S., Mituyama, T., Kawamura, Y., Ono, Y., Sakota, E., Kotani, H., Asai, K., Siomi, H. and Siomi, M.C. (2009) A regulatory circuit for piwi by the large Maf gene traffic jam in Drosophila. Nature **461**, 1296–1301
28. Aravin, A.A., Sachidanandam, R., Bourc'his, D., Schaefer, C., Pezic, D., Toth, K.F., Bestor, T. and Hannon, G.J. (2008) A piRNA pathway primed by individual transposons is linked to de novo DNA methylation in mice. Mol. Cell **31**, 785–799
29. Rouget, C., Papin, C., Boureux, A., Meunier, A.C., Franco, B., Robine, N., Lai, E.C., Pelisson, A. and Simonelig, M. (2011) Maternal mRNA deadenylation and decay by the piRNA pathway in the early Drosophila embryo. Nature **467**, 1128–1132
30. Li, M.A., Alls, J.D., Avancini, R.M., Koo, K. and Godt, D. (2003) The large Maf factor Traffic Jam controls gonad morphogenesis in Drosophila. Nat. Cell Biol. **5**, 994–1000
31. Saito, K., Ishizu, H., Komai, M., Kotani, H., Kawamura, Y., Nishida, K.M., Siomi, H. and Siomi, M.C. (2010) Roles for the Yb body components Armitage and Yb in primary piRNA biogenesis in Drosophila. Genes Dev. **24**, 2493–2498
32. Haase, A.D., Fenoglio, S., Muerdter, F., Guzzardo, P.M., Czech, B., Pappin, D.J., Chen, C., Gordon, A. and Hannon, G.J. (2010) Probing the initiation and effector phases of the somatic piRNA pathway in Drosophila. Genes Dev. **24**, 2499–2504
33. Olivieri, D., Sykora, M.M., Sachidanandam, R., Mechtler, K. and Brennecke, J. (2010) An in vivo RNAi assay identifies major genetic and cellular requirements for primary piRNA biogenesis in Drosophila. EMBO J. **29**, 3301–3317
34. Qi, H., Watanabe, T., Ku, H.Y., Liu, N., Zhong, M. and Lin, H. (2011) The Yb body, a major site for Piwi associated RNA biogenesis and a gateway for Piwi expression and transport to the nucleus in somatic cells. J. Biol. Chem. **286**, 3789–3797
35. Nishimasu, H., Ishizu, H., Saito, K., Fukuhara, S., Kamatani, M.K., Matsumoto, N., Nishizawa, T., Nakanaga, K., Aoki, J. et al. (2012). Structure and function of Zucchini endoribonuclease in piRNA biogenesis. Nature **491**, 284–287
36. Klattenhoff, C., Xi, H., Li, C., Lee, S., Xu, J., Khurana, J.S., Zhang, F., Schultz, N., Koppetsch, B.S., Nowosielska, A. et al. (2009) The Drosophila HP1 homolog Rhino is required for transposon silencing and piRNA production by dual-strand clusters. Cell **138**, 1137–1149
37. Pane, A., Jiang, P., Zhao, D.Y., Singh, M. and Schüpbach, T. (2011) The Cutoff protein regulates piRNA cluster expression and piRNA production in the Drosophila germline. EMBO J. **30**, 4601–4615
38. Lim, A. K and Kai, T. (2007) Unique germ-line organelle, nuage, functions to repress selfish genetic elements in Drosophila melanogaster. Proc. Natl Acad. Sci. U.S.A. **104**, 6714–6719
39. Rangan, P., Malone, C.D., Navarro, C., Newbold, S.P., Hayes, P.S., Sachidanandam, R., Hannon, G.J. and Lehmann, R. (2011) piRNA production requires heterochromatin formation in Drosophila. Curr. Biol. **21**,1373–1379
40. Anand, A. and Kai, T. (2011) The tudor domain protein kumo is required to assemble the nuage and to generate germline piRNAs in Drosophila. EMBO J. **31**, 870–882
41. Liu, L., Qi, H., Wang, J. and Lin, H. (2011) PAPI, a novel TUDOR-domain protein, complexes with AGO3, ME31B and TRAL in the nuage to silence transposition. Development **138**, 1863–1873
42. Zhang, Z., Xu, J., Koppetsch, B.S., Wang, J., Tipping, C., Ma, S., Weng, Z., Theurkauf, W.E. and Zamore, P.D. (2011) Heterotypic piRNA Ping-Pong requires qin, a protein with both E3 ligase and Tudor domains. Mol. Cell **44**, 572–584
43. Patil, V.S. and Kai, T. (2010) Repression of retroelements in Drosophila germline via piRNA pathway by the Tudor domain protein Tejas. Curr. Biol. **20**, 724–730
44. Nishida, K.M., Okada, T.N., Kawamura, T., Mituyama, T., Kawamura, Y., Inagaki, S., Huang, H., Chen, D., Kodama, T., Siomi, H. and Siomi, M.C. (2009) Functional involvement of Tudor and dPRMT5 in the piRNA processing pathway in Drosophila germlines. EMBO J. **28**, 3820–3831

45. Handler, D., Olivieri, D., Novatchkova, M., Gruber, F.S., Meixner, K., Mechtler, K., Stark, A., Sachidanandam, R. and Brennecke, J. (2011) A systematic analysis of *Drosophila* TUDOR domain-containing proteins identifies Vreteno and the Tdrd12 family as essential primary piRNA pathway factors. EMBO J. **30**, 3977–3993
46. Zamparini, A.L., Davis, M.Y., Malone, C.D., Vieira, E., Zavadil, J., Sachidanandam, R., Hannon, G.J. and Lehmann, R. (2011) Vreteno, a gonad-specific protein, is essential for germline development and primary piRNA biogenesis in *Drosophila*. Development **138**, 4039–4050
47. Chen, C., Nott, T.J., Jin, J. and Pawson, T. (2011) Deciphering arginine methylation: Tudor tells the tale. Nat. Rev. Mol. Cell Biol. **12**, 629–642
48. Findley, S.D., Tamanaha, M., Clegg, N.J. and Ruohola-Baker, H. (2003) Maelstrom, a *Drosophila* spindle-class gene, encodes a protein that colocalizes with Vasa and RDE1/AGO1 homolog, Aubergine, in nuage. Development **130**, 859–871
49. Sato, K., Nishida, K.M., Shibuya, A., Siomi, M.C. and Siomi, H. (2011) Maelstrom co-ordinates microtubule organization during *Drosophila* oogenesis through interaction with components of the MTOC. Genes Dev. **25**, 2361–2373
50. Pane, A., Wehr, K. and Schüpbach, T. (2007) zucchini and squash encode two putative nucleases required for rasiRNA production in the *Drosophila* germline. Dev. Cell **12**, 851–862
51. Sato, K. and Siomi, H. (2010) Is canalization more than just a beautiful idea? Genome Biol. **11**, 109–111
52. Specchia, V., Piacentini, L., Tritto, P., Fanti, L., D'Alessandro, R., Palumbo, G., Pimpinelli, S. and Bozzetti, M.P. (2010) Hsp90 prevents phenotypic variation by suppressing the mutagenic activity of transposons. Nature **463**, 662–665
53. Gangaraju, V.K., Yin, H., Weiner, M.M., Wang, J., Huang, X.A. and Lin, H. (2011) *Drosophila* Piwi functions in Hsp90-mediated suppression of phenotypic variation. Nat. Genet. **43**, 153–158
54. Esteller M. (2011) Non-coding RNAs in human disease. Nat. Rev. Genet. **12**, 861–874
55. Siddiqi, S. and Matushansky, I. (2012) Piwis and piwi-interacting RNAs in the epigenetics of cancer. J. Cell Biochem. **113**, 373–380
56. Janic, A., Mendizabal, L., Llamazares, S., Rossell, D. and Gonzalez, C. (2010) Ectopic expression of germline genes drives malignant brain tumor growth in *Drosophila*. Science **330**, 1824–1827
57. Nagao, A., Sato, K., Nishida, K.M., Siomi, H. and Siomi, M.C. (2011) Gender-specific hierarchy in nuage localization of PIWI-interacting RNA factors in *Drosophila*. Front. Genet. **2**, 55
58. Anne, J. and Mechler, B.M. (2005) Valois, a component of the nuage and pole plasm, is involved in assembly of these structures, and binds to Tudor and the methyltransferase Capsuléen. Development **132**, 2167–2177

Small nucleolar RNAs and RNA-guided post-transcriptional modification

Lauren Lui and Todd Lowe[1]

Department of Biomolecular Engineering, University of California Santa Cruz, Santa Cruz, CA 95064, U.S.A.

Abstract

snoRNAs (small nucleolar RNAs) constitute one of the largest and best-studied classes of non-coding RNAs that confer enzymatic specificity. With associated proteins, these snoRNAs form ribonucleoprotein complexes that can direct 2′-O-methylation or pseudouridylation of target non-coding RNAs. Aided by computational methods and high-throughput sequencing, new studies have expanded the diversity of known snoRNA functions. Complexes incorporating snoRNAs have dynamic specificity, and include diverse roles in RNA silencing, telomerase maintenance and regulation of alternative splicing. Evidence that dysregulation of snoRNAs can cause human disease, including cancer, indicates that the full scope of snoRNA roles remains an unfinished story. The diversity in structure, genomic origin and function between snoRNAs found in different complexes and among different phyla illustrates the surprising plasticity of snoRNAs in evolution. The ability of snoRNAs to direct highly specific interactions with other RNAs is a consistent thread in their newly discovered functions. Because they are ubiquitous throughout Eukarya and Archaea, it is likely they were a feature of the last common ancestor of these two domains, placing their origin over two billion years ago. In the present chapter, we focus on recent advances in our understanding of these ancient, but functionally dynamic RNA-processing machines.

Keywords:
2′-O-methylation, post-transcriptional modification, pseudouridylation, ribosome biogenesis, small nucleolar RNA.

[1]To whom correspondence should be addressed (email lowe@soe.ucsc.edu).

Introduction

Although snoRNAs (small nucleolar RNAs) are still primarily characterized by their ability to direct post-transcriptional modification of RNAs, recent research has revealed diverse functions, expression strategies and genomic organization. Excellent reviews exist on snoRNAs that detail their best-understood molecular roles [1–8]. The present chapter seeks to highlight new discoveries and the functional elasticity of this class of ncRNA (non-coding RNA). snoRNAs have had a long history of revealing unexpected new functions – scientists did not characterize their defining role in post-transcriptional modification until three decades after their discovery. First observed in the nucleolus of mammalian cells in the late 1960s and early 1970s ([9,10], reviewed in [11]), snoRNAs were not yet clearly divided into the two major families we understand today. Research in the late 1980s and early 1990s confirmed earlier predicted roles for these RNAs in directing pre-ribosomal RNA cleavage ([12–14], reviewed in [1,8]). Later in the mid-1990s, scientists used conserved sequence features to divide snoRNAs into the families that are currently accepted today [15,16]. In 1996, two separate studies demonstrated that 'C/D box' RNAs can direct 2′-O-methylation (Figures 1A and 1B) of rRNA (ribosomal RNA) [17,18]. A year later, two groups also demonstrated that 'H/ACA box' RNAs direct pseudouridylation (Figures 1C and 1D) of rRNA [19,20]. Follow-up studies showed that the vast majority of C/D box RNAs in *Saccharomyces cerevisiae* are indeed required for almost all 2′-O-methyl modifications in yeast rRNA [21].

In 2000, biochemical and computational experiments surprisingly confirmed the existence of C/D box sRNAs (sno-like RNAs) in Archaea, the 'third domain' of life [22,23]. Later experiments confirmed *in vitro* methylation activity of target sites on rRNA and tRNA [24,25]. Although archaeal organisms are prokaryotic, and therefore lack a nucleus and nucleolus, the presence of snoRNA machinery contributes to the body of evidence indicating that archaea are more closely related to eukaryotes than bacteria. Bacteria also have pseudouridylated and 2′-O-methylated RNAs, but do not have snoRNAs. Instead, they use site- or region-specific protein enzymes to modify ribonucleotides [26]. Archaea and eukaryotes also have similar DNA replication, transcription and translation systems compared with those in bacteria. The existence of C/D box and H/ACA box RNAs, along with their associated orthologous proteins in archaea and eukaryotes, suggests that this system of RNA modification is an ancient one, estimated at over two billion years old [27–29]. New evidence indicates snoRNAs have evolved additional, somewhat unexpected, cellular roles since their origin.

Studies in the past decade linking snoRNA expression to cancer highlight how incomplete our understanding remains. In 2002, Chang et al. [30] provided evidence that an H/ACA box RNA is significantly down-regulated in human brain cancer, the first report of a relationship between snoRNA dysregulation and cancer. Since then, dysregulation of snoRNAs has been linked to lymphoma, breast, prostate, lung, and head and neck cancers (reviewed in [31,32]). Some snoRNAs have tumour suppressor roles, whereas the up-regulation of others may contribute to tumorigenesis. Mutations of snoRNA-associated proteins or snoRNAs that modify rRNA can also cause ribosome dysfunction, leading to a variety of cancers and other diseases [33,34]. Studies of snoRNAs and snoRNA host genes linked to cancer suggest that snoRNA host genes, previously thought to have no protein-coding value, may indeed have roles in cell development and homoeostasis (reviewed in [32]). Because some snoRNAs have differential expression patterns in specific cancer types, the levels of particular snoRNAs in

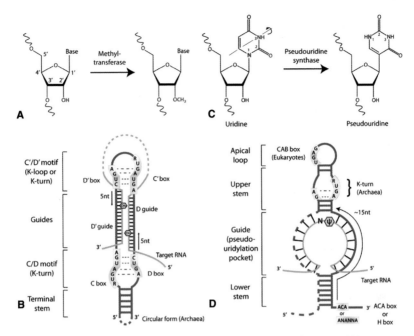

Figure 1. Sequence and structural features of C/D and H/ACA box RNAs
(**A**) Addition of a 2′-O-methyl to a ribonucleotide is indicated in red. This type of modification is associated with C/D box RNAs. (**B**) Secondary structure of a typical C/D box RNA. Conserved box elements (yellow) pair to form kink-turns (black broken lines). The C′/D′ pair form an apical loop kink-turn known as the K-loop. The D and D′ guides of the C/D box RNA (blue) are complementary sequences to a target RNA (orange). Both guides may target sites on the same RNA. The methyltransferase fibrillarin adds a 2′-O-methyl (CH_3, green) to the nucleotide (nt) of the target RNA that is base-paired to the guide RNA 5 nt upstream of the D or D′ box. Circular forms of C/D box RNAs exist in some archaeal species, such as *Pyrococcus furiosus* [115,116]. (**C**) Conversion of a uridine into a pseudouridine by isomerization. This type of modification is associated with H/ACA box RNAs. (**D**) Secondary structure of a typical H/ACA box RNA. Conserved box elements are shown in yellow. H/ACA box RNAs can be composed of one, two or three hairpins. Each hairpin is comprised of a lower stem, a bulge known as the pseudouridylation pocket, an upper stem and an apical loop. Recently it was discovered that the 5′-end of the hairpin up to the upper stem can be missing [46]. Hairpins are separated by the H box (ANANNA) and the last hairpin contains an ACA box at its 3′-end. In archaea, a kink-turn can form in the upper stem. The pseudouridylation pocket contains sequence antisense to the target RNA, analogous to the guides of the C/D box RNA. The uridine to be isomerized to pseudouridine (ψ, green) is located under the upper stem, approximately 15 nt upstream of the ACA or H box. H/ACA box RNAs and scaRNAs contain a Cajal body (CAB) box, which localizes the guide RNA to Cajal bodies to modify small nuclear RNAs.

the blood could be used as a non-invasive method to detect and type specific cancers [31]. Cancer research is expanding our current understanding of how snoRNAs and snoRNA host genes affect cell state, further underlining new roles to be studied.

Although the focus of the present chapter is to highlight new functions of snoRNAs, the study of snoRNAs has been driven by the importance of 2′-O-methylation and pseudouridylation in non-coding RNAs. 2′-O-Methylation and pseudouridylation are the most commonly found RNA modifications and play important roles in the stability and folding of RNA [26,35]. Pseudouridines are so numerous they have been called the 'fifth nucleoside' [36]. Over 100 sites in human rRNA are targeted for methylation, and nearly that many are pseudouridylated [26,37]. Most snoRNAs target rRNAs, but in eukaryotes these RNAs also target snRNAs (small nuclear RNAs) involved with the spliceosome [38] and tRNAs in archaea [39,40]. Modifications by snoRNAs often cluster around functionally important regions in rRNA, snRNAs and tRNAs, suggesting that modification plays a role in the function of these RNAs (reviewed in [2,41]). RNA modification by snoRNAs represents a fine-tuning mechanism for a molecule that is initially built with only four different units.

To fully understand the emerging functional roles of snoRNAs, potential synthetic applications, and the promise of accelerated detection of new snoRNA genes by computational methods, it is helpful to review their biogenesis and defining sequence features. We focus mainly on the largest subset of C/D box RNAs and H/ACA box RNAs which direct the 2'-O-methylation and pseudouridylation of target RNAs respectively [17–20].

Composition and structure of snoRNPs

As with many RNA families, both primary sequence and structural features play important roles in snoRNA function. snoRNAs serve as scaffolds in the formation of the snoRNP (small nucleolar ribonucleoprotein) complex, as well as acting as base-pairing guides to target specific ribonucleotides for modification. As computational biologists, we closely observe the natural variation in conserved sequence and secondary structural features, as this enables the development of sensitive computational models to detect this class of RNAs and their targets.

C/D box RNPs

C/D box RNAs are ~50–300 nt in length and contain characteristic fairly well-conserved C (RUGAGA) and D (CUGA) box elements, and similar but less conserved C' and D' box elements (Figure 1B, see [4]). Archaeal C/D box sRNAs are shorter than their eukaryotic cousins (~50–70 nt), but also contain C, D, C' and D' boxes with remarkably similar sequence motifs as their eukaryotic counterparts. The C and D boxes, located respectively at the 5' and 3' termini of the RNA, form the 'kink-turn', a stem-bulge-stem RNA structure that is critical to the formation of the snoRNP [42]. L7Ae (archaea) or the 15.5 kDa protein (eukaryotes) binds to the kink-turn, allowing the rest of the proteins associated with the C/D box RNP to bind. In archaeal species, the internal C' and D' boxes also form a 'K-loop', a version of the kink-turn with one stem replaced by a loop, and which is also bound by L7Ae [42]. The conserved C and D box features and formation of the kink-turn are hallmarks of this class of snoRNA.

Typically, eukaryotic C/D box RNAs contain a 10–21 nt guide sequence complementary to the target RNA to be modified by the methyltransferase fibrillarin. This guide sequence is typically shorter in archaeal C/D box sRNAs, generally ranging from 8 to 12 nt in length. Budding yeast snoRNAs tend to be longer than those in most other studied species, with additional sequence in the region between the C' and D' boxes. Recently, a study of yeast C/D box RNAs found that many of these extra 'middle region' sequences have additional complementarities with their target RNAs. These additional sites of interaction flank the target methylation site, and can stimulate methylation efficiency up to 5-fold [43].

An evolutionarily well-conserved set of core proteins associate with C/D box RNAs to form a unique RNA-directed molecular machine with the striking ability to add 2'-O-methyl groups to a theoretically unbounded number of different, but highly specific, positions across any number of RNA targets (reviewed in [44]). The range of specificity of this enzymatic complex is only limited by the number of different snoRNAs encoded within the genome. In eukarya, the 15.5 kDa snoRNP protein binds to the kink-turn formed by the C and D boxes. Next, Nop56, Nop58 and fibrillarin bind to complete the RNP. In archaea, the snoRNP is simpler, comprised of the C/D box RNA bound to L7Ae (a homologue with the 15.5 kDa protein), Nop 56/58 (related to both eukaryotic Nop56 and Nop58), and fibrillarin. Although these core proteins are clearly homologous between Archaea and Eukarya, the assembly of the RNP differs between the two domains. Recent research on the assembly of C/D box RNPs is discussed in further detail below.

H/ACA box RNPs

RNAs that belong to the H/ACA box RNA family share hallmark secondary structure and primary sequence elements that are quite unique from C/D box snoRNAs. Typical H/ACA box RNAs are ~60–150 nt in length (longer in yeast) and are named for a hinge sequence with a consensus of ANANNA and an ACA element at the 3′-end (Figure 1D) [15,19]. The terminal ACA motif is something of a misnomer in some species, as it can commonly vary, leading it to also be referred to as the 'ANA' terminal motif. For example, in trypanosomes, the ACA can be AGA [45], and in some unusual archaeal thermophiles, the ACA can be omitted altogether [46]. Structurally, H/ACA box RNAs are often comprised of two hairpins separated by the hinge element, but the number of hairpins can vary. Eukaryotic H/ACA box RNAs have two hairpins, except in trypansomes where they can be singlets [47]. In yeast, both hairpin structures are needed for modification even if the H/ACA box RNA only has one pseudouridylation target [48]. Single hairpins have also been found in the protist *Euglena gracilis* [49]. H/ACA box RNAs with one, two or three hairpins have been found in archaea [50,51].

Each hairpin is comprised of a lower stem, a central bulge dubbed the 'pseudouridylation pocket', an upper stem and the apical loop. The pseudouridylation pocket is analogous to the guide sequence of C/D box RNAs, but is discontinuous with the 'left guide' and 'right guide' regions that are interrupted by the upper stem and apical loop. Within the context of its tertiary structure, the left and right guides are in close proximity, yet the separation in the linear sequence of the two complementary regions makes computational detection of H/ACA box RNA genes and their targets much more difficult. When the left and right guide sequences within the pseudouridylation pocket base pair with the target RNA, two nucleotides of the target RNA remain unpaired at the base of the upper stem. The 5′-most nucleotide of this pair is converted from uridine into pseudouridine. Often in archaea, the upper stem forms a kink-turn that is recognized by L7Ae. In eukaryotes this kink-turn is not known to form, but NHP2, the eukaryotic homologue of L7Ae, is structurally similar to L7Ae and may interact with the upper stem of H/ACA box RNAs [52]. Despite the conservation of secondary structure across species, some H/ACA box RNAs in the archaeal genus *Pyrobaculum* are missing the lower stem [46], suggesting that minimal forms of H/ACA box RNAs could exist undetected in other species as well.

The four core proteins of H/ACA box RNPs are Cbf5 (also known as dyskerin/DKC1 in mammals or NAP57), Nop10, Gar1 and L7Ae (Figure 2). The key protein in the complex, Cbf5, is a pseudouridine synthase which catalyses the isomerization of the target uridine to pseudouridine. It contains a catalytic domain and a PUA (pseudouridine and archeosine transglycosylase) domain. The PUA domain has a dual role as an RNA-binding motif and a localization signal to the nucleus (reviewed [36]). Mutations of this domain are linked to dyskeratosis congenita,

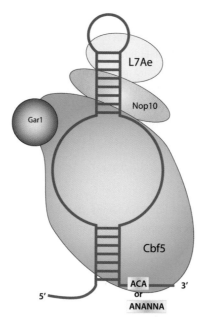

Figure 2. H/ACA box RNP
The core proteins of the H/ACA box RNP are Cbf5, L7Ae, Nop10 and Gar1. Similar to how the 15.5 kDa protein takes the place of L7Ae in eukaryotic C/D box RNPs, L7Ae is replaced with NHP2 in eukaryotic H/ACA box RNPs. Unlike the 15.5 kDa protein, NHP2 is not known to have kink-turn binding properties although it is a homologue of L7Ae. See Table 1 for H/ACA box RNP protein homologues between eukaryotes and archaea.

a rare inherited bone marrow disease [4]. The ACA box and lower stem of H/ACA box RNAs are bound by the PUA domain, which positions the guide regions by the catalytic domain. Studies of Cbf5 alone and complexed in the H/ACA box RNP indicate that it is structurally similar to TruB, the bacterial pseudouridine synthase which does not harness an RNA cofactor for specificity [6]. In some species, Cbf5 isomerizes uridines of tRNAs without the presence of other H/ACA box RNA proteins [53,54], which also also supports its common ancestry with TruB.

scaRNPs (small Cajal body-specific RNAs)

scaRNAs are a related hybrid type of snoRNA that associate with C/D box and H/ACA box RNA proteins, and direct both 2′-O-methylation and pseudouridylation of snRNAs [55]. These RNAs contain CAB boxes (Cajal body box, consensus UGAG) in the apical loop of the hairpin. CAB boxes localize the scaRNA to Cajal bodies, nuclear organelles that function in snRNP maturation [3].

Functions of snoRNAs beyond ribosomal processing and modification

Although the majority of C/D box and H/ACA box RNAs known today are annotated for their roles in RNA modification, early studies recognized and detailed their essential role in rRNA processing during ribosome biogenesis (reviewed in [1]). Among eukaryotes, uridine-rich snoRNAs such as U3, U14, U22, snR10 and snR30 act as integral components of the processing machinery by guiding endonucleolytic cleavage of pre-rRNA and folding of the mature rRNA (reviewed in [8,56]). In both eukaryotes and archaea, the proximal base-pairing sites between 2′-O-methyl guide RNAs and rRNA further support their general chaperone role in rRNA folding [8,40]. Recent studies also indicate that RNA helicases regulate the association of snoRNPs with rRNA [8].

There is now ample evidence that other classes of RNAs interact directly with snoRNAs. For years, snoRNA studies have noted C/D box or H/ACA box snoRNAs that have potential guide sequences with no clear modification targets among rRNAs or other known modification targets (spliceosomal RNAs in eukaryotes [38]; tRNAs in archaea [39,40]). The presence of these 'orphan' guide snoRNA genes in most species suggests there are additional target RNAs to be identified [57,58]. Computational search algorithms, expression screens and high-throughput sequencing have greatly aided traditional RNA biochemical approaches in identifying these new interactions.

snoRNAs and RNA silencing

Studies indicate that some snoRNAs are processed into miRNAs (microRNAs), another important and incompletely understood class of small ncRNAs. First discovered in 1993 [59], miRNAs are ~18–24 nt long RNAs that associate with RNA silencing machinery to regulate the stability and translation of messenger RNAs [60]. The development of high-throughput sequencing methods for small RNAs has accelerated our appreciation of the complexity and dynamics of small RNA populations, including snoRNAs and miRNAs.

In many cases, miRNAs derived from snoRNAs (termed sno-miRNAs or sno-derived RNAs) were discovered by a combination of bioinformatics and analysis of deep sequencing of RNAs physically associated with core proteins of the RNA silencing pathway, such as Argonaute [61,62]. sno-miRNAs derived from both C/D and H/ACA box snoRNAs have been found in a variety of eukaryotes, including human, mouse, fruit fly, plants and fission yeast [61–65]. C/D box snoRNAs that act as miRNA precursors have also been found in the parasitic protozoan *Giardia lamblia* [66] and the Epstein–Barr virus [67]. Sequencing libraries that have been enriched for small RNAs by size fractionation also aid the search for snoRNAs and sno-miRNAs [65,68]. Determining snoRNA database entries that overlap with miRNAs has also yielded new discoveries [65].

How snoRNAs are processed into sno-miRNAs is still unclear, but the mechanism may include traditional snoRNA processing machinery [7] and RNAi (RNA interference) proteins [61]. Some sno-miRNAs retain the traditional functionality of snoRNAs and are localized in the nucleolus [64,69]. H/ACA box snoRNAs that act as miRNA precursors have been shown to bind to snoRNA-associated proteins, including dyskerin [64] and GAR-1 [62]. Populations of sno-derived small RNAs are also affected by the loss of components of the RNAi pathway, such as Dicer [62].

This new role of snoRNAs in gene silencing may partially explain the presence of 'orphan guides' in some cases. The targets of these conserved guides may be mRNAs to be silenced, instead of rRNAs to be modified. Most of the RNAs derived from snoRNAs that have been studied to date appear to have miRNA-like functions, but derived RNAs longer than 22 nt may have other functions [7]. The emerging relationship between snoRNAs and RNA silencing may reveal more about the evolution of RNA in gene regulation.

© 2013 Biochemical Society

Telomerase: a specialized H/ACA box RNA in vertebrates

Telomeres, the ends of eukaryotic chromosomes, are maintained by the RNP complex telomerase which consists of a protein reverse transcriptase [TERT (telomerase reverse transcriptase)] and TR (telomerase RNA; also abbreviated to TER or TERC). Without telomerase, telomeres would shorten and eventually be lost due to incomplete replication of telomere ends by conventional DNA polymerases [70]. In vertebrates, the TR is approximately 450 nt long and has H/ACA box RNA features at its 3′-end which are required for proper localization and accumulation. One of these features, the CAB box, directs telomerase to Cajal bodies [3]. Vertebrate TR interacts with all of the core proteins of H/ACA box RNAs, but interestingly, there is no evidence of a pseudouridylation target [3]. Mutations in TR or its associated proteins can lead to the syndrome dyskeratosis congenita [4].

Spliced leader mRNAs in trypanosomes

Maturation of all mRNAs in trypanosomes requires the incorporation of a 39 nt spliced leader exon that is added *in trans* [47,71]. A trypanosome H/ACA box RNA has been shown to guide pseudouridylation of the spliced leader RNA [47], providing an important example of mRNA modification by snoRNAs, rather than the typical rRNA, tRNA and snRNA targets.

Artificial uses of snoRNAs

Scientists have utilized the main features of snoRNAs, such as the ability to target specific RNA sequences and localization to the nucleolus, for synthetic biology purposes. C/D box RNAs have been used to regulate gene expression [72,73], regulate alternative splicing [74,75], localize ribozymes to the nucleolus in yeast [76] and guide site-specific modification [77]. Modified H/ACA box RNAs have also been used to convert translation termination codons into sense codons to suppress translation termination. Not only does this method suggest that mRNAs may be naturally modified, it indicates a potential way to treat premature termination codon diseases, such as cystic fibrosis and Duchenne muscular dystrophy [78,79]. The ability to synthesize C/D box RNAs that affect alternative splicing and gene expression by only changing guide sequences [72,75], together with current stories of non-canonical snoRNA function [7,80] suggest that natural snoRNAs could have already evolved these functions. High-throughput sequencing technologies to detect post-transcriptional modifications in mRNAs is rapidly improving, which should aid in searches for novel mRNA–snoRNA interactions.

An unexpected pathological role of snoRNAs in human disease

In the late 1990s and early in the new millennium, it was found that dysregulation of snoRNA expression can cause human disease [81,82]. The best understood example is the genetic disorder PWS (Prader–Willi syndrome), a congenital disease characterized by mental retardation, hyperphagia leading to obesity and short stature. It affects approximately 1 in 8000–20000 people and is the most common genetic cause for Type II diabetes [80,83].

Investigation of the genetic defects associated with PWS led to the discovery of tissue-specific snoRNAs in the brain, and implicates processed snoRNAs in the regulation of alternative splicing. Minimal deletions in the SNURF-SNRPN locus on paternal chromosome 5q11-q13 were genetically linked to PWS [84]. The SNURF-SNRPN pre-mRNA contains a heterogenous cluster of intronic orphan snoRNAs, and is maternally imprinted (only expressed from the paternal allele). This atypical snoRNA gene cluster includes multiple copies of SNORD115 (also known as HBII-52), which only occurs in the brain [85]. SNORD115 is processed into shorter forms that can bind to the alternative exon 5b of the serotonin-2C receptor mRNA [80]. The hybridization of the snoRNA with the pre-mRNA masks a splicing silencer, allowing for the incorporation of the alternative exon in the mature mRNA [80,85]. Bioinformatics analysis also indicates that the mouse form of the SNORD115 (MBII-52) may regulate alternative splicing of mRNAs other than that of the serotonin-2C receptor [80]. Yet another snoRNA in the SNURF-SNRPN locus, SNORD116/ HBII-85, may regulate alternative splicing of other mRNA targets [86]. The continuing studies of PWS suggest that there are undiscovered functional roles of snoRNAs outside of RNA modification.

Diversity in expression and organization among species

The varied genomic organization of C/D and H/ACA box RNAs within individual genomes and across different phyla illustrates the dynamic transcriptional units and processing pathways associated with this RNA family. snoRNA genes may occur as independently transcribed genes, within introns of protein-coding genes or lncRNAs (long ncRNAs), or in clusters that are transcribed as a polycistronic transcript (reviewed in [4,5,87]).

All species studied have at least some fraction of their snoRNAs driven by independent promoters, either as monocistrons or polycistronic transcripts (Figure 3). Most eukaryotic snoRNAs that fall into this class are transcribed by RNA Pol II (RNA polymerase II). The few that are transcribed by RNA Pol III have additional transcription elements such as internal box A and box B motifs, or the upstream sequence element in plants. Also in plants are discistronic tRNA-snoRNA genes that are transcribed by RNA Pol III from the tRNA promoter [5]. C/D box RNAs processed from polycistrons are trimmed by exonucleases (reviewed in [4,7]).

With the exception of protozoans, most sequenced eukaryotic genomes contain intron-encoded snoRNAs. In a few notable instances, snoRNAs are found in the introns of lncRNAs in mammals [88], and in select archaea there is an instance of a C/D box sRNA encoded within the intron of a tRNA gene [39]. The snoRNAs located in introns can occur singly or in clusters. These clusters can be homogenous (copies of same snoRNA gene) or heterogenous; in some cases, introns may even be composed of both C/D and H/ACA box RNAs. Some of the most highly conserved snoRNAs across phyla are commonly found in core genes such as those encoding ribosomal proteins (e.g. RPL7A), but in other cases, host genes do not appear to have a significant protein product (e.g. GAS5) [4].

snoRNAs found within polycistrons or introns require processing by nucleases before they are functional (reviewed in [7]). In yeast, polycistronic snoRNAs are cleaved preferentially in hairpin structures containing AGNN tetraloops by Rnt1p, an RNase III homologue. Extra sequence is cleaved away by $5' \rightarrow 3'$ and $3' \rightarrow 5'$ exonucleases (Table 1). Intronic snoRNAs can be processed from introns released as lariats or, more rarely, excised by endonucleases. Intronic snoRNAs originating from lariat-structured introns are also trimmed by exonucleases. In

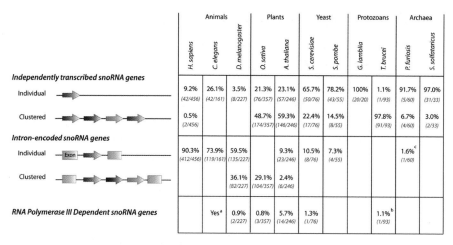

Figure 3. Types of genomic organization of snoRNA genes with distribution within species
Animal, plant, and yeast numbers are based on data collected by [5], *G. lamblia* data from [117], *T. brucei* data from [118] and archaeal data from [22,23,40]. MRP RNA genes are not included in this table. The diversity of snoRNAs is illustrated by the variety of genomic locations in which they are found. snoRNA genes (coloured block arrows) can be independently transcribed with their own promoters or found in the introns of protein and (rarely) RNA genes. Clusters may be composed of both C/D and H/ACA box RNAs, such as in plants and trypanosomes [47]. Typically mammalian host genes are involved in ribosomal biogenesis or protein synthesis [98], but several do not have a significant protein product. snoRNAs independently transcribed or encoded within introns are transcribed by RNA Pol II unless otherwise indicated. [a]For *C. elegans*, the number of snoRNA coding units is reported based on data from [5]. These units have elements indicating that they are potentially recognized by RNA Pol III and are already included in the other categories of this table based on their genomic location. [b]Only the U3 snoRNA homologue is known to be independently transcribed by RNA Pol III in trypanosomes [119]. [c]Some archaeal species that belong to the phylum Euryarchaeota, such as *Pyrococcus furiosus*, have a C/D box RNA encoded within tRNA-TRP [25].

eukaryotes, snoRNAs mature in Cajal bodies, although the nucleolus is the final destination of most snoRNAs.

Evidence supports the propagation of snoRNAs by transposable elements and other types of duplications. In plants, the large presence of snoRNAs is attributed to polyploidy, large chromosomal rearrangements and tandem duplications (reviewed in [87,89]). Bioinformatics analyses indicate that mammalian snoRNAs result from retrotransposition events [90,91]. In the human genome, hundreds of snoRNAs and snoRNA-related molecules appear to be derived from transposons. These snoRNAs have features of retrotransposons, such as an A-rich tail and are flanked by a target site duplication sequence that corresponds to their insertion site. Diversity of genomic location provides a challenge to the computational prediction methods used to find and annotate snoRNAs, in addition to variation in sequence and structure.

Computational search methods and databases

Computational methods can greatly accelerate the search for plausible snoRNA candidates relative to traditional biochemical isolation experiments, but a fair number of challenges remain.

Table 1. Processing, assembly and core proteins of snoRNPs

HUGO gene IDs for vertebrate proteins are included after the gene name for reference. Note that in archaea snoRNA processing proteins are not known. The Table is modified from [7], with additions from [8].

Protein	Vertebrate	Yeast	Archaea	Function
Core C/D box RNP proteins	15.5 kDa (*NHP2L1*)	Snu13p	L7Ae	Bind to kink-turn formed by C/D motifs (and C′/D′ motif in archaea), nucleation of snoRNP assembly
	Fibrillarin (*FBL*)	Nop1p	Fibrillarin	Methyltransferase
	Nop56 (*NOP56*)	Nop56p	Nop56/58	Recruits fibrillarin
	Nop58 (*NOP58*)	Nop58p/ Nop5p		
Core H/ACA box RNP proteins	NHP2 (*NHP2*)	Nhp2p	L7Ae	Bind to kink-turn (archaea), nucleation of snoRNP assembly
	NAP57/dyskerin (*DKC1*)	Cbf5p	Cbf5	Pseudouridine synthase
	GAR1 (*GAR1*)	Gar1p	Gar1	Target RNA turnover
	NOP10 (*NOP10*)	Nop10p	Nop10	Interaction with L7Ae and core RNA
Assembly proteins	Naf1 (*NAF1*)	Naf1p		Shuttle protein, accumulation of H/ACA box RNAs, placeholder for GAR1
	Shq1 (*SHQ1*)	Shq1p		Shuttle protein, placeholder for H/ACA box RNA
	Bcd1 (*KLF6*)	Bcd1		Accumulation of C/D box RNAs, coupling of transcription and processing of snoRNA
	IBP160 (*AQR*)			Coupling between pre-mRNA splicing and snoRNA assembly
	TIP48 (*RUVBL2*); TIP49 (*RUVBL1*)	TIP48p; TIP49p	TIP49-like protein	Bridge interactions between the 15.5 kDa and the NOP56 and NOP58 proteins
	NOP17 (*PIH1D1*)			C/D box RNA assembly

(*Continued*)

Table 1. Processing, assembly and core proteins of snoRNPs (*Continued*)

Protein	Vertebrate	Yeast	Archaea	Function
	TAF9 (*TAF9*)			C/D box RNA assembly
	NUFIP (*NUFIP1*)			Regulation of TIP48 and TIP49 interactions with core C/D box RNA proteins
	Hsp90 (*HSP90AA1*)			Assists folding of NHP2 and 15.5 kDa proteins
RNA-processing proteins	Xrn1 (*XRN1*); Xrn2 (*XRN2*)	Xrn1p; Rat1p/Xrn2p		5′ → 3′ exonuclease trimming
	Rrp6 (*EXOSC10*); Rrp46 (*EXOSC5*)	Rrp6p; Rrp46p		3′ → 5′ exonuclease trimming
		Rnt1p		Excision of snoRNA from hairpin (RNase III activity)
	Lsm; La (*SBB*)	Lsmp; Lhp1p		Stabilization of mature 3′-end of snoRNA
	TGS1 (*TGS1*)			Hypermethylates m^{7}G-gap of snoRNAs from independent genes
		Prp43; Has1p		Release/association of snoRNAs with rRNA (helicase activity)
Localization proteins	PHAX (*PHAX*)			Transport of C/D box RNAs to Cajal bodies
	CRM1 (*XPO1*)	CRM1		Transport of C/D box RNAs from Cajal bodies to nucleoli
	Ran (*RAN*)			snoRNA trafficking, nuclear export
	CBP20 (*NCBP2*); CBP80 (*NCBP1*)			Cap-binding complex proteins
	Nopp140 (*NOLC1*)	Srp40p		Nucleocytoplasmic transport factor

Despite conservation of some sequence and structural features, the variation of snoRNAs in structure, length and genomic context across different phyla has hampered efforts to create general universally effective computational search methods for the two major types of snoRNA. On the basis of the presence of associated proteins in all eukaryotic and archaeal species, and careful study of a small number of model species, we estimate that the majority of snoRNA genes have not been identified or annotated properly for most species in public databases [92]. BLAST [93], a popular sequence similarity search algorithm, is poor at identifying homologues because of the rapid divergence of guide sequences [40] and the relatively short length of the conserved sequence features. Although a portion of methylation sites are widely conserved, many are conserved only between closely related species or within phyla [40]. Recent discoveries that stretch the possibilities of functional H/ACA and C/D box RNA structure [46,94,95] illustrate the need for computational programs that are specialized by clades (Table 2).

The recent availability of high-throughput sequencing data strengthens computational searches by narrowing 'search space' and confirming predictions with transcriptional data. The increasing availability of sequenced genomes also enables the use of comparative genome analysis in the search for snoRNAs [23,64] and ncRNAs in general [96]. RNA sequencing libraries that are enriched for small RNAs has greatly assisted bioinformatics searches for H/ACA and C/D box RNAs [46,51,68,97]. Some existing programs increased specificity by determining antisense targets of guide regions [21,98], but miss orphan guides. With transcriptional data, we can confirm the presence of predicted snoRNAs with unknown targets [68]. In the archaeal species *Pyrobaculum aerophilum*, the number of C/D box snoRNAs was increased ~37% (65–89 genes) by using transcriptional data and eliminating the requirement for antisense RNA

Table 2. Computational snoRNA prediction methods and databases

Program or database	Description	Website	Reference
snoTarget	Mammalian C/D box RNA predictor	Server: http://bpg.utoledo.edu/~dbs/snotarget/	[86]
		Standalone: http://www.utoledo.edu/med/depts/bioinfo/snotarget.html	
snoReport	C/D and H/ACA box RNA predictor	Standalone: http://www.bioinf.uni-leipzig.de/~jana/index.php/jana-hertel-software/64-jana-hertel-snoreport	[120]
snoscan	C/D box RNA predictor	Server and standalone: http://lowelab.ucsc.edu/snoscan/	[21]
snoGPS	Human, yeast, archaeal H/ACA RNA predictor	Server and standalone: http://lowelab.ucsc.edu/snoGPS/	[121]

(*Continued*)

Table 2. Computational snoRNA prediction methods and databases (*Continued*)

Program or database	Description	Website	Reference
snoSeeker	Mammalian C/D and H/ACA box RNA predictor	Server and standalone: http://genelab.sysu.edu.cn/snoseeker/index.php	[122]
ERPIN and RNAMOT	Archaeal H/ACA RNA prediction method	Method outlined in reference	[123]
		ERPIN: http://tagc.univ-mrs.fr/erpin	
		RNAMOT: http://pages-perso.esil.univ-mrs.fr/~dgaut/download/	
RNAsnoop	H/ACA box RNA target predictor	Standalone: http://www.bioinf.uni-leipzig.de/~htafer/RNAsnoop/RNAsnoop.html	[124]
PLEXY	C/D box RNA target predictor	Standalone: http://www.bioinf.uni-leipzig.de/Software/PLEXY	[125]
Psiscan	Trypanosome H/ACA box RNA predictor	Method described in reference	[126]
Plant snoRNA Database	Plant snoRNA sequences and target sites	http://bioinf.scri.sari.ac.uk/cgi-bin/plant_snorna/home	[127]
Yeast snoRNA Database at UMass-Amherst	Yeast snoRNA sequences and target sites	http://people.biochem.umass.edu/fournierlab/snornadb/main.php	[128]
Rfam Database	Archaeal and eukaryotic snoRNA sequences and families	http://rfam.sanger.ac.uk/	[102]
snoRNABase	Human snoRNA sequence and target sites	http://www-snorna.biotoul.fr/	[37]
Lowe/Eddy snoRNA Database	Yeast, archaea and *Arabidopsis thaliana* C/D box RNA sequences and alignments	http://lowelab.ucsc.edu/snoRNAdb/	[21,23,127]

targets [68], adding substantially to a list of carefully curated predictions that had existed for 10 years [99].

A variety of snoRNA databases exist, with most entries originally (or only) identified by bioinformatics screens. The lack of a common nomenclature can lead to confusion when comparing snoRNA homologues and pseudogenes [100], although the nomenclature SNORD, SNORA and SCARNA has been developed for human C/D box RNAs, H/ACA box RNAs and scaRNAs respectively [101]. Many of these databases also include information about predicted target sites. One notable resource, the Rfam database [102] does not annotate predicted targets of snoRNAs due to its general RNA search methodology [103], but it acts as a valuable repository of most classes of RNA, including specific snoRNA genes that are conserved between three or more species.

snoRNP biogenesis

snoRNAs form the nucleating backbone of snoRNPs, to which associating proteins bind to form the mature RNP. Study of the assembly of these complexes aids our understanding of diseases affected by snoRNPs and rRNA biogenesis. In addition, the differences between the archaeal and eukaryotic versions reveal subtle changes in the evolutionary interplay between the RNAs and their associated proteins.

C/D box RNPs

In archaea, the conventional model of C/D box RNPs is 'symmetrical', and involves one snoRNA molecule per complex (Figure 4A; reviewed in [44]). First, the L7Ae protein initiates assembly by binding to the kink-turn formed by the C and D boxes, followed by a second L7Ae protein binding to the K-loop formed by the C′ and D′ boxes. Next, two Nop56/58 (also known as Nop5) proteins bind, one at each pole of the snoRNA. Two fibrillarin proteins then complete the symmetrical assembly with a total of six proteins bound to a single RNA. Interestingly, loss of L7Ae K-loop binding does not abolish its incorporation into the C′/D′ portion of the RNP, nor does it destroy methylation of the target nucleotide of the D′-guide. Likely protein–protein interactions are an important factor in C/D box RNP assembly, not just initiation by L7Ae [104].

Follow-up studies of the archaeal snoRNP have offered a new alternative model. In 2009, Bleichert et al. [105] proposed a dimeric C/D box RNP structure based on EM (electron microscopy) data of an archaeal C/D sRNP (Figure 4B). This 'di-sRNP' consists of two C/D box RNAs and four of each core protein, contrary to the conventional model, and can be extended to the eukaryotic model (Figure 4D). As a counterpoint to the di-sRNP model, a 2011 study by Lin et al. [106] synthesized two RNAs that could form kink-turn structures and act as a scaffold for Nop5, L7Ae and fibrillarin [106]. This study resulted in a crystal structure of a monomeric RNP that also had proven *in vitro* activity. However, the structure in the Lin et al. [106] study did not contain a K-loop between the C′ and D′ boxes, which is typical of archaeal C/D box RNAs and thus may call into question the applicability of the results. To investigate the importance of the K-loop, Bower-Phipps and Taylor [107] demonstrated in 2012 that nearly all C/D box sRNPs with an internal loop in their C/D sRNA do adopt the di-sRNP architecture. By testing C/D box sRNAs from several widely divergent archaeal species, this study concluded that the di-sRNP structure is likely to be conserved across the archaeal domain. It remains unclear if computationally predicted (but functionally unverified) C/D box sRNAs that lack a K-loop could potentially assemble naturally into the monomeric RNP form.

Less is known about eukaryotic C/D box RNP structure, but it appears to be 'asymmetric' since the 15.5 kDa protein only binds to the kink-turn formed by the terminal C and D boxes (Figure 4C). On the basis of *in vivo* cross-linking experiments in oocytes of the African clawed frog, *Xenopus laevis*, Nop56 is predicted to bind to the C′/D′ motifs and Nop58 to the C/D motifs [108]. Fibrillarin is predicted to bind both sets of motifs. In contrast, a study in the yeast *S. cerevisiae* indicates that all four core proteins associate with both the C/D and C′/D′ motifs, on the basis of results of co-immunoprecipitation experiments [109]. This work

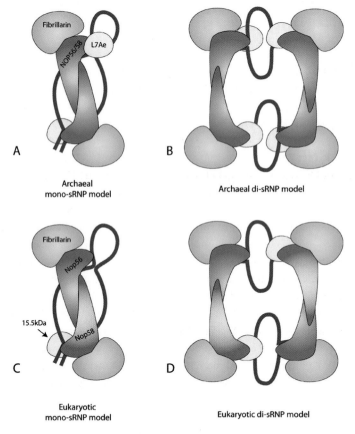

Figure 4. C/D box sRNP models
Fully assembled C/D box sRNPs differ between Archaea and Eukarya, and various models exist based on crystal structures and EM data. The model illustrated here is based on Figures in [129]. L7Ae/15.5 kDa, yellow; Nop56/58 and Nop56, light blue; Nop58, purple; fibrillarin, green; and C/D box RNA, dark blue. (**A**) Conventional 'symmetrical' archaeal C/D box RNP model supported by crystallographic data [106]. (**B**) Proposed archaeal di-sRNP model. The di-sRNP model is based on an EM-reconstruction of archaeal C/D box RNPs from [107]. (**C**) Conventional 'asymmetrical' eukaryotic C/D box RNP model. This architecture is supported by biochemical data [108,130]. In the 'asymmetrical' eukaryotic model, Nop58 binds to the C/D motifs and Nop56 to the C′/D′ motifs and the 15.5 kDa protein (homologous with L7Ae) only binds to the kink-turn formed by the C and D boxes. (**D**) Proposed eukaryotic di-sRNP model.

demonstrated that the individual C/D and C′/D′ RNPs are coupled spatially and functionally for methylation activity [109]. The association of the 15.5 kDa protein with the C′/D′ motif, despite its inability to bind the K-loop formed by the motif, suggests the possibility that the eukaryotic C/D box RNP structure is symmetrical, like the archaeal version. This interaction echoes the ability of archaeal L7Ae to associate with the C′/D′ motif even if it is mutated so it can no longer bind to the K-loop [104]. Similar to archaeal C/D box RNP assembly, protein–protein interactions play a large role in the formation of the snoRNP complex.

In eukaryotes, a growing number of proteins have been implicated in the biogenesis of C/D box RNPs (see Table 1). Among them are the proteins IBP160 and Bcd1 (box C/D RNA 1) which are implicated as assembly factors [4]. IBP160 is a general splicing factor that binds upstream of intronic C/D box RNAs and starts the biogenesis of the C/D box RNP. Bcd1 is necessary for the accumulation of C/D box RNAs, but not a member of the final active C/D box RNP. The role of other identified assembly and localization factors such as CRM1 is unclear and the subject of current research (reviewed in [8]).

H/ACA box RNPs

Owing to differences between archaeal L7Ae and the eukaryotic NHP2, the assembly of archaeal and eukaryotic H/ACA box RNPs differs (reviewed in [36]). NHP2 belongs to the same protein family as L7Ae, but does not bind kink-turns like its archaeal homologues. For the archaeal complex, L7Ae and Cbf5 bind directly to the guide RNA. Nop10 and Gar1 bind to Cbf5 independently, forming a stable heterotrimeric complex. In contrast in eukaryotes, NHP2 does not bind RNA directly, and must be recruited to the H/ACA box RNP by protein–protein interactions, forming a complex with dyskerin (homologue with Cbf5) and Nop10. This protein complex binds to the hairpin of the H/ACA box RNA. Finally, Gar1, named for the GAR (glycine-arginine rich) domains flanking the central domain of the protein, binds to dyskerin.

Compared with archaeal H/ACA box RNP assembly, eukaryotic H/ACA box RNP assembly is complex, requiring multiple assembly factors (reviewed in [8]). Naf1 (nuclear assembly factor 1) appears to be involved in the biogenesis of H/ACA box RNPs, but is not necessary for the active RNP, where Gar1 takes the place of Naf1. Gar1 contains a C-terminal extension that regulates substrate turnover and allows Cbf5 to bind tighter than Naf1 [110]. Naf1 is needed for the accumulation of H/ACA box RNPs, so the exchange of Naf1 for Gar1 may be a key step in regulating the activity of this complex. Like Naf1, Shq1 is a nucleoplasmic shuttle protein and may act as a placeholder for the H/ACA box RNA during RNP assembly [111].

Evolution of L7Ae and kink-turn-binding proteins

snoRNPs of both eukaryotes and archaea contain kink-turn-binding proteins or homologues that belong to the L7Ae/L30e protein family. The RNA kink-turn motif is characterized by stacked sheared GA base pairs flanked by two stems, or by a stem and a loop in the case of K-loops. This motif acts as the binding site for L7Ae and its homologues (e.g. 15.5 kDa protein), nucleating the assembly of the RNP. In the case of eukaryotic H/ACA box RNPs, NHP2 belongs to the L7Ae/L30e protein family, but does not appear to have binding specificity to kink-turns.

The existence of only two kink-turn-binding proteins in archaea and six in eukaryotes indicates diversification of context and/or function across the tree of life [44]. Other eukaryotic members of this family occur in specialized RNPs; for example, the SBP2 protein is required for selenocysteine incorporation, mediated by binding a kink-turn structure in the 3′ UTR (untranslated region) of selenoprotein mRNAs. Recently, YbxF and YlxQ were demonstrated to be bacterial homologues of L7Ae [112]. These proteins bind kink-turns, but not as strongly as L7Ae, and do not bind K-loops. In eukaryotes, studies suggest that Hsp (heat-shock protein) 90 works in concert with conserved cofactors to control the biogenesis of RNPs with kink-turn-binding proteins [113]. L7Ae in archaea has also unexpectedly been found to be part of the tRNA-processing RNase P complex [114]. These findings invite broader questions about possible regulatory cross-talk between diverse RNA processing pathways, including snoRNPs, which appear to compete for a common component.

Conclusions

Previously thought to have a singular role in ribosome biogenesis, we now have diverse evidence that snoRNAs play a broader role in the cell. This is exemplified by cases of dysregulation of these RNAs that have unanticipated effects on the organism. Protein translation, pre-mRNA splicing, telomere stability and thus cell viability depend on the function of snoRNAs. The various roles of snoRNAs in translation suggest that snoRNAs may have originated from an early translation system [29]. Future elucidation of the versatility of snoRNA function will advance from a mix of biochemical, computational and high-throughput sequencing approaches.

Summary

- Guide C/D box RNPs and H/ACA box RNPs are defined by their ability to carry out precise 2′-O-methylation and pseudouridylation of target nucleotides.
- The assembly of snoRNPs (small nucleolar RNAs) is complex and differs in some aspects between the eukaryotic and archaeal domains.
- snoRNAs are transcribed from a variety of genomic contexts, illustrating evolutionary diversity and maturation pathways.
- snoRNAs have functions and roles beyond post-transcriptional modification, such as ribosomal RNA cleavage, regulation of gene expression by RNA silencing and alternative splicing, and telomerase maintenance.
- Disruption and mutations to snoRNPs can cause human diseases.
- Computational search methods exist for snoRNAs, but the diversity in sequence motifs strongly suggests that other forms of snoRNAs probably exist that current methods do not detect.

References

1. Maxwell, E. and Fournier, M.J. (1995) The small nucleolar RNAs. Annu. Rev. Biochem. **35**, 897–934
2. Yu, Y., Terns, R.M. and Terns, M.P. (2005) Mechanisms and functions of RNA-guided RNA. In Topics in Current Genetics, Volume 12 (Grosjean, H., ed.), pp. 223–262, Springer-Verlag, Berlin Heidelberg

3. Terns, M. and Terns, R. (2006) Noncoding RNAs of the H/ACA family. Cold Spring Harbor Symp. Quant. Biol. **71**, 395–405
4. Matera, G., Terns, R.M. and Terns, M.P. (2007) Non-coding RNAs: lessons from the small nuclear and small nucleolar RNAs. Nat. Rev. Mol. Cell Biol. **8**, 209–220
5. Dieci, G., Preti, M. and Montanini, B. (2009) Eukaryotic snoRNAs: a paradigm for gene expression flexibility. Genomics **94**, 83–88
6. Kiss, T., Fayet-Lebaron, E. and Jády, B.E. (2010) Box H/ACA small ribonucleoproteins. Mol. Cell **37**, 597–606
7. Falaleeva, M. and Stamm, S. (2012) Fragments of small nucleolar RNAs as a new source for noncoding RNAs. In Regulatory RNAs (Mallick, B. and Ghosh, Z., eds), pp. 49–71, Springer-Verlag, Berlin, Heidelberg
8. Watkins, N.J. and Bohnsack, M.T. (2012) The box C/D and H/ACA snoRNPs: key players in the modification, processing and the dynamic folding of ribosomal RNA. WIREs RNA **3**, 397–414
9. Weinberg, R.A. and Penman, S. (1968) Small molecular weight monodisperse nuclear RNA. J. Mol. Biol. **38**, 289–304
10. Nakamura, T., Prestayko, A. and Busch, H. (1968) Studies on nucleolar 4 to 6 S ribonucleic acid of Novikoff hepatoma cells. J. Biol. Chem. **243**, 1368–1375
11. Reddy, R. and Busch, H. (1988) Small nuclear RNAs: RNA sequences, structure, and modifications. In Structure and Function of Major and Minor Small Nuclear Ribonucleoprotein Particles (Birnstiel, M.L., ed.), pp. 1–37, Springer-Verlag, New York
12. Kass, S., Tyc, K., Steitz, J.A. and Sollner-Webb, B. (1990) The U3 small nucleolar ribonucleoprotein functions in the first step of preribosomal RNA processing. Cell **60**, 897–908
13. Tollervey, D. (1987) A yeast small nuclear RNA is required for normal processing of pre-ribosomal RNA. EMBO J. **6**, 4169–4175
14. Beltrame, M. and Tollervey, D. (1995) Base pairing between U3 and the pre-ribosomal RNA is required for 18S rRNA synthesis. EMBO J. **14**, 4350–4356
15. Balakin, A.G., Smith, L. and Fournier, M.J. (1996) The RNA world of the nucleolus: two major families of small RNAs defined by different box elements with related functions. Cell **86**, 823–834
16. Tollervey, D. and Kiss, T. (1997) Function and synthesis of small nucleolar RNAs. Curr. Opin. Cell Biol. **9**, 337–342
17. Kiss-László, Z., Henry, Y., Bachellerie, J.P., Caizergues-Ferrer, M., and Kiss, T. (1996) Site-specific ribose methylation of preribosomal RNA: a novel function for small nucleolar RNAs. Cell **85**, 1077–1088
18. Nicoloso, M., Qu, L.H., Michot, B., and Bachellerie, J.P. (1996) Intron-encoded, antisense small nucleolar RNAs: the characterization of nine novel species points to their direct role as guides for the 2′-O-ribose methylation of rRNAs. J. Mol. Biol. **260**, 178–195
19. Ganot, P., Bortolin, M.L. and Kiss, T. (1997) Site-specific pseudouridine formation in preribosomal RNA is guided by small nucleolar RNAs. Cell **89**, 799–809
20. Ni, J., Tien, A.L., and Fournier, M.J. (1997) Small nucleolar RNAs direct site-specific synthesis of pseudouridine in ribosomal RNA. Cell **89**, 565–573
21. Lowe, T.M. and Eddy, S.R. (1999) A computational screen for methylation guide snoRNAs in yeast. Science **283**, 1168–1171
22. Gaspin, C., Cavaillé, J., Erauso, G., and Bachellerie, J.P. (2000) Archaeal homologs of eukaryotic methylation guide small nucleolar RNAs: lessons from the *Pyrococcus* genomes. J. Mol. Biol. **297**, 895–906
23. Omer, A.D., Lowe, T.M., Russell, G., Ebhardt, H., Eddy, S.R. and Dennis, P.P. (2000) Homologs of small nucleolar RNAs in Archaea. Science **288**, 517–522
24. Omer, A.D., Ziesche, S., Ebhardt, H., and Dennis, P.P. (2002) *In vitro* reconstitution and activity of a C/D box methylation guide ribonucleoprotein complex. Proc. Natl. Acad. Sci. U.S.A. **99**, 5289–5294

25. Clouet d'Orval, B., Bortolin, M.L., Gaspin, C. and Bachellerie, J.P. (2001) Box C/D RNA guides for the ribose methylation of archaeal tRNAs. The tRNATrp intron guides the formation of two ribose-methylated nucleosides in the mature tRNATrp. Nucleic Acids Res. **29**, 4518–4529
26. Decatur, W.A. and Fournier, M.J. (2002) rRNA modifications and ribosome function. Trends Biochem. Sci. **27**, 344–351
27. Brocks, J.J. (1999) Archean molecular fossils and the early rise of eukaryotes. Science **285**, 1033–1036
28. Alvarez-Ponce, D. and McInerney, J.O. (2011) The human genome retains relics of its prokaryotic ancestry: human genes of archaebacterial and eubacterial origin exhibit remarkable differences. Genome Biol. Evol. **3**, 782–790
29. Tran, E., Brown, J. and Maxwell, E.S. (2004) Evolutionary origins of the RNA-guided nucleotide-modification complexes: from the primitive translation apparatus? Trends Biochem. Sci. **29**, 343–350
30. Chang, L. Sen, Lin, S.Y., Lieu, A.S. and Wu, T.L. (2002) Differential expression of human 5S snoRNA genes. Biochem. Biophys. Res. Commun. **299**, 196–200
31. Mannoor, K., Liao, J. and Jiang, F. (2012) Small nucleolar RNAs in cancer. Biochim. Biophys. Acta **1826**, 121–128
32. Williams, G.T. and Farzaneh, F. (2012) Are snoRNAs and snoRNA host genes new players in cancer? Nat. Rev. Cancer **12**, 84–88
33. Higa-Nakamine, S., Suzuki, T., Uechi, T., Chakraborty, A., Nakajima, Y., Nakamura, M., Hirano, N., Suzuki, T. and Kenmochi, N. (2012) Loss of ribosomal RNA modification causes developmental defects in zebrafish. Nucleic Acids Res. **40**, 391–398
34. Heiss, N., Knight, S. and Vulliamy, T. (1998) X-linked dyskeratosis congenita is caused by mutations in a highly conserved gene with putative nucleolar functions. Nat Genet. **19**, 32–38
35. Helm, M. (2006) Post-transcriptional nucleotide modification and alternative folding of RNA. Nucleic Acids Res. **34**, 721–733
36. Grozdanov, P.N. and Meier, U.T. (2009) Multicomponent machines in RNA modification: H/ACA ribonucleoproteins. In DNA and RNA Modification Enzymes: Structure, Mechanism, Function, and Evolution, (Grosjean, H., ed.), pp. 450–460, Landes Bioscience, Austin, TX
37. Lestrade, L. and Weber, M.J. (2006) snoRNA-LBME-db, a comprehensive database of human H/ACA and C/D box snoRNAs. Nucleic Acids Res. **34**, D158–D162
38. Tycowski, K.T., You, Z.H., Graham, P.J. and Steitz, J.A. (1998) Modification of U6 spliceosomal RNA is guided by other small RNAs. Mol. Cell **2**, 629–638
39. Singh, S.K., Gurha, P., Tran, E.J., Maxwell, E.S. and Gupta, R. (2004) Sequential 2′-O-methylation of archaeal pre-tRNATrp nucleotides is guided by the intron-encoded but trans-acting box C/D ribonucleoprotein of pre-tRNA. J. Biol. Chem. **279**, 47661–47671
40. Dennis, P.P., Omer, A., and Lowe, T. (2001) A guided tour: small RNA function in Archaea. Mol. Microbiol. **40**, 509–519
41. Durairaj, A. and Limbach, P.A. (2008) Mass spectrometry of the fifth nucleoside: a review of the identification of pseudouridine in nucleic acids. Anal. Chim. Acta **623**, 117–125
42. Nolivos, S., Carpousis, A.J. and Clouet-d'Orval, B. (2005) The K-loop, a general feature of the *Pyrococcus* C/D guide RNAs, is an RNA structural motif related to the K-turn. Nucleic Acids Res. **33**, 6507–6514
43. van Nues, R.W., Granneman, S., Kudla, G., Sloan, K.E., Chicken, M., Tollervey, D. and Watkins, N.J. (2011) Box C/D snoRNP catalysed methylation is aided by additional pre-rRNA base-pairing. EMBO J. **30**, 2420–2430
44. Gagnon, K.T., Qu, G., and Maxwell, E.S. (2009) Multicomponent 2′-O-ribose methylation machines: evolving box C/D RNP structure and function. In DNA and RNA Modification Enzymes: Structure, Mechanism, Function, and Evolution, (Grosjean, H., ed.), pp. 436–449, Landes Bioscience, Austin, TX
45. Barth, S., Shalem, B., Hury, A., Tkacz, I.D., Liang, X.H., Uliel, S., Myslyuk, I., Doniger, T., Salmon-Divon, M., Unger, R. and Michaeli, S. (2008) Elucidating the role of C/D snoRNA in rRNA processing and modification in *Trypanosoma brucei*. Eukaryotic Cell **7**, 86–101

46. Bernick, D.L., Dennis, P.P., Höchsmann, M. and Lowe, T.M. (2012) Discovery of Pyrobaculum small RNA families with atypical pseudouridine guide RNA features. RNA **18**, 402–411
47. Uliel, S., Liang, X., Unger, R. and Michaeli, S. (2004) Small nucleolar RNAs that guide modification in trypanosomatids: repertoire, targets, genome organisation, and unique functions. Int. J. Parasitol. **34**, 445–454
48. Bortolin, M.L., Ganot, P. and Kiss, T. (1999) Elements essential for accumulation and function of small nucleolar RNAs directing site-specific pseudouridylation of ribosomal RNAs. EMBO J. **18**, 457–469
49. Russell, A.G., Schnare, M.N. and Gray, M.W. (2004) Pseudouridine-guide RNAs and other Cbf5p-associated RNAs in *Euglena gracilis*. RNA **10**, 1034–1046
50. Muller, S., Leclerc, F., Behm-Ansmant, I., Fourmann, J.B., Charpentier, B. and Branlant, C. (2008) Combined *in silico* and experimental identification of the *Pyrococcus abyssi* H/ACA sRNAs and their target sites in ribosomal RNAs. Nucleic Acids Res. **36**, 2459–2475
51. Tang, T.-H., Bachellerie, J.-P., Rozhdestvensky, T., Bortolin, M.L., Huber, H., Drungowski, M., Elge, T., Brosius, J. and Hüttenhofer, A. (2002) Identification of 86 candidates for small non-messenger RNAs from the archaeon *Archaeoglobus fulgidus*. Proc. Natl. Acad. Sci. U.S.A. **99**, 7536–7541
52. Koo, B.-K., Park, C.-J., Fernandez, C.F., Chim, N., Ding, Y., Chanfreau, G. and Feigon, J. (2011) Structure of H/ACA RNP protein Nhp2p reveals *cis/trans* isomerization of a conserved proline at the RNA and Nop10 binding interface. J. Mol. Biol. **411**, 927–942
53. Muller, S., Fourmann, J.-B., Loegler, C., Charpentier, B. and Branlant, C. (2007) Identification of determinants in the protein partners aCBF5 and aNOP10 necessary for the tRNA:Psi55-synthase and RNA-guided RNA:Psi-synthase activities. Nucleic Acids Res. **35**, 5610–5624
54. Roovers, M., Hale, C., Tricot, C., Terns, M.P., Terns, R.M., Grosjean, H. and Droogmans, L. (2006) Formation of the conserved pseudouridine at position 55 in archaeal tRNA. Nucleic Acids Res. **34**, 4293–4301
55. Darzacq, X., Jády, B.E., Verheggen, C., Kiss, A.M., Bertrand, E. and Kiss, T. (2002) Cajal body-specific small nuclear RNAs: a novel class of 2′-O-methylation and pseudouridylation guide RNAs. EMBO J. **21**, 2746–2756
56. Henras, A.K., Soudet, J., Gérus, M., Lebaron, S., Caizergues-Ferrer, M., Mougin, A. and Henry, Y. (2008) The post-transcriptional steps of eukaryotic ribosome biogenesis. Cell. Mol. Life Sci. **65**, 2334–2359
57. Jády, B.E. and Kiss, T. (2000) Characterisation of the U83 and U84 small nucleolar RNAs: two novel 2′-O-ribose methylation guide RNAs that lack complementarities to ribosomal RNAs. Nucleic Acids Res. **28**, 1348–1354
58. Bachellerie, J.P., Cavaillé, J. and Hüttenhofer, A. (2002) The expanding snoRNA world. Biochimie **84**, 775–790
59. Lee, R.C., Feinbaum, R.L. and Ambros, V. (1993) The *C. elegans* heterochronic gene lin-4 encodes small RNAs with antisense complementarity to lin-14. Cell **75**, 843–854
60. Winter, J., Jung, S., Keller, S., Gregory, R.I. and Diederichs, S. (2009) Many roads to maturity: microRNA biogenesis pathways and their regulation. Nat. Cell Biol. **11**, 228–234
61. Taft, R.J., Glazov, E.A., Lassmann, T., Hayashizaki, Y., Carninci, P. and Mattick, J.S. (2009) Small RNAs derived from snoRNAs. RNA **15**, 1233–1240
62. Ender, C., Krek, A., Friedländer, M.R., Beitzinger, M., Weinmann, L., Chen, W., Pfeffer, S., Rajewsky, N. and Meister, G. (2008) A human snoRNA with microRNA-like functions. Mol. Cell **32**, 519–528
63. Cole, C., Sobala, A., Lu, C. and Thatcher, S. (2009) Filtering of deep sequencing data reveals the existence of abundant Dicer-dependent small RNAs derived from tRNAs. RNA **15**, 2147–2160
64. Scott, M.S., Avolio, F., Ono, M., Lamond, A.I. and Barton, G.J. (2009) Human miRNA precursors with box H/ACA snoRNA features. PLoS Comput. Biol. **5**, e1000507

65. Brameier, M., Herwig, A., Reinhardt, R., Walter, L. and Gruber, J. (2011) Human box C/D snoRNAs with miRNA like functions: expanding the range of regulatory RNAs. Nucleic Acids Res. **39**, 675–686
66. Saraiya, A.A. and Wang, C.C. (2008) snoRNA, a novel precursor of microRNA in *Giardia lamblia*. PLoS Pathog. **4**, e1000224
67. Hutzinger, R., Feederle, R., Mrazek, J., Schiefermeier, N., Balwierz, P.J., Zavolan, M., Polacek, N., Delecluse, H.J. and Hüttenhofer, A. (2009) Expression and processing of a small nucleolar RNA from the Epstein–Barr virus genome. PLoS Pathog. **5**, e1000547
68. Bernick, D.L., Dennis, P.P., Lui, L.M. and Lowe, T.M. (2012) Diversity of antisense and other non-coding RNAs in archaea revealed by comparative small RNA sequencing in four *Pyrobaculum* species. Front. Microbiol. **3**, 231
69. Politz, J.C.R., Hogan, E.M. and Pederson, T. (2009) MicroRNAs with a nucleolar location. RNA **15**, 1705–1715
70. Lingner, J., Cooper, J.P. and Cech, T.R. (2012) Telomerase end replication: lagging strand and no DNA longer problem? Science **269**, 1533–1534
71. Zamudio, J.R., Mittra, B., Chattopadhyay, A., Wohlschlegel, J.A., Sturm, N.R. and Campbell, D.A. (2009) *Trypanosoma brucei* spliced leader RNA maturation by the cap 1 2′-O-ribose methyltransferase and SLA1 H/ACA snoRNA pseudouridine synthase complex. Mol. Cell. Biol. **29**, 1202–1211
72. Ono, M., Yamada, K., Avolio, F., Scott, M.S., van Koningsbruggen, S., Barton, G.J. and Lamond, A.I. (2010) Analysis of human small nucleolar RNAs (snoRNA) and the development of snoRNA modulator of gene expression vectors. Mol. Biol. Cell **21**, 1569–1584
73. Saito, H., Kobayashi, T., Hara, T., Fujita, Y., Hayashi, K., Furushima, R. and Inoue, T. (2010) Synthetic translational regulation by an L7Ae-kink-turn RNP switch. Nat. Chem. Biol. **6**, 71–78
74. Zhao, X. and Yu, Y. (2007) Targeted pre-mRNA modification for gene silencing and regulation. Nat. Methods **5**, 95–100
75. Stepanov, G.A., Semenov, D.V, Kuligina, E.V., Koval, O.A., Rabinov, I.V., Kit, Y.Y. and Richter, V.A. (2012) Analogues of artificial human box C/D small nucleolar RNA as regulators of alternative splicing of a pre-mRNA target. Acta Nat. **4**, 32–41
76. Samarsky, D.A. and Fournier, M.J. (1999) A comprehensive database for the small nucleolar RNAs from *Saccharomyces cerevisiae*. Nucleic Acids Res. **27**, 161–164
77. Liu, B., Ni, J. and Fournier, M.J. (2001) Probing RNA *in vivo* with methylation guide small nucleolar RNAs. Methods **23**, 276–286
78. Karijolich, J. and Yu, Y.-T. (2011) Converting nonsense codons into sense codons by targeted pseudouridylation. Nature **474**, 395–398
79. Huang, C., Wu, G. and Yu, Y.-T. (2012) Inducing nonsense suppression by targeted pseudouridylation. Nat. Protoc. **7**, 789–800
80. Kishore, S., Khanna, A., Zhang, Z., Hui, J., Balwierz, P.J., Stefan, M., Beach, C., Nicholls, R.D., Zavolan, M. and Stamm, S. (2010) The snoRNA MBII-52 (SNORD 115) is processed into smaller RNAs and regulates alternative splicing. Hum. Mol. Genet. **19**, 1153–1164
81. Cavaillé, J., Buiting, K., Kiefmann, M., Lalande, M., Brannan, C.I., Horsthemke, B., Bachellerie, J.P., Brosius, J. and Hüttenhofer, A. (2000) Identification of brain-specific and imprinted small nucleolar RNA genes exhibiting an unusual genomic organization. Proc. Natl. Acad. Sci. U.S.A. **97**, 14311–14316
82. Runte, M., Hüttenhofer, A., Gross, S., Kiefmann, M., Horsthemke, B. and Buiting, K. (2001) The IC-SNURF-SNRPN transcript serves as a host for multiple small nucleolar RNA species and as an antisense RNA for UBE3A. Hum. Mol. Genet. **10**, 2687–2700
83. Nagai, T. and Mori, M. (1999) Prader-Willi Syndrome, diabetes mellitus and hypogonadism. Biomed. Pharmacother. **53**, 452–454
84. Sahoo, T., Del Gaudio, D., German, J.R., Shinawi, M., Peters, S.U., Person, R.E., Garnica, A., Cheung, S.W. and Beaudet, A.L. (2008) Prader-Willi phenotype caused by paternal deficiency for the HBII-85 C/D box small nucleolar RNA cluster. Nat. Genet. **40**, 719–721

85. Kishore, S. and Stamm, S. (2006) Regulation of alternative splicing by snoRNAs. Cold Spring Harbor Symp. Quant. Biol. **71**, 329–334
86. Bazeley, P.S., Shepelev, V., Talebizadeh, Z., Butler, M.G., Fedorova, L., Filatov, V. and Fedorov, A. (2008) snoTARGET shows that human orphan snoRNA targets locate close to alternative splice junctions. Gene **408**, 172–179
87. Rodor, J., Letelier, I., Holuigue, L. and Echeverria, M. (2010) Nucleolar RNPs: from genes to functional snoRNAs in plants. Biochem. Soc. Trans. **38**, 672–676
88. Le Meur, E., Watrin, F., Landers, M., Sturny, R., Lalande, M. and Muscatelli, F. (2005) Dynamic developmental regulation of the large non-coding RNA associated with the mouse 7C imprinted chromosomal region. Dev. Biol. **286**, 587–600
89. Brown, J.W.S., Echeverria, M. and Qu, L.H. (2003) Plant snoRNAs: functional evolution and new modes of gene expression. Trends Plant Sci. **8**, 42–49
90. Weber, M.J. (2006) Mammalian small nucleolar RNAs are mobile genetic elements. PLoS Genet. **2**, e205
91. Luo, Y. and Li, S. (2007) Genome-wide analyses of retrogenes derived from the human box H/ACA snoRNAs. Nucleic Acids Res. **35**, 559–571
92. Gardner, P.P., Bateman, A. and Poole, A.M. (2010) SnoPatrol: how many snoRNA genes are there? J. Biol. **9**, 4
93. Altschul, S., Gish, W., Miller, W., Myers, E. and Lipman, D. (1990) Basic local alignment search tool. J. Mol. Biol. **215**, 403–410
94. Blaby, I.K., Majumder, M., Chatterjee, K., Jana, S., Grosjean, H., de Crécy-Lagard, V. and Gupta, R. (2011) Pseudouridine formation in archaeal RNAs: the case of *Haloferax volcanii*. RNA **17**, 1367–1380
95. Joardar, A., Malliahgari, S. R., Skariah, G. and Gupta, R. (2011) 2′-O-methylation of the wobble residue of elongator pre-tRNAmet in *Haloferax volcanii* is guided by a box C/D RNA containing unique features. RNA Biol. **8**, 1–10
96. Washietl, S., Hofacker, I.L. and Stadler, P.F. (2005) Fast and reliable prediction of noncoding RNAs. Proc. Natl. Acad. Sci. U.S.A. **102**, 2454–2459
97. Chen, H.-M. and Wu, S.-H. (2009) Mining small RNA sequencing data: a new approach to identify small nucleolar RNAs in *Arabidopsis*. Nucleic Acids Res. **37**, e69
98. Söderbom, F. (2006) Small nucleolar RNAs: identification, structure, and function. In Nucleic Acids and Molecular Biology, Volume 17, Small RNAs: Analysis and Regulatory Functions (Nellen, W. and Hammann, C., eds), pp. 31–56, Springer-Verlag, Berlin Heidelberg
99. Fitz-Gibbon, S.T., Ladner, H., Kim, U.-J., Stetter, K.O., Simon, M.I. and Miller, J.H. (2002) Genome sequence of the hyperthermophilic crenarchaeon *Pyrobaculum aerophilum*. Proc. Natl. Acad. Sci. U.S.A. **99**, 984–989
100. Makarova, J.A. and Kramerov, D.A. (2011) SNOntology: myriads of novel snoRNAs or just a mirage? BMC Genomics **12**, 543
101. Bruford, E.A., Lush, M.J., Wright, M.W., Sneddon, T.P., Povey, S. and Birney, E. (2008) The HGNC database in 2008: a resource for the human genome. Nucleic Acids Res. **36**, D445–D448
102. Gardner, P.P., Daub, J., Tate, J., Moore, B.L., Osuch, I.H., Griffiths-Jones, S., Finn, R.D., Nawrocki, E.P., Kolbe, D.L., Eddy, S.R. and Bateman, A. (2011) Rfam: Wikipedia, clans and the 'decimal' release. Nucleic Acids Res. **39**, D141–D145
103. Nawrocki, E.P., Kolbe, D.L. and Eddy, S.R. (2009) Infernal 1.0: inference of RNA alignments. Bioinformatics **25**, 1335–1337
104. Gagnon, K.T., Zhang, X., Qu, G., Biswas, S., Suryadi, J., Brown, 2nd, B.A. and Maxwell, E.S. (2010) Signature amino acids enable the archaeal L7Ae box C/D RNP core protein to recognize and bind the K-loop RNA motif. RNA **16**, 79
105. Bleichert, F., Gagnon, K.T., Brown, 2nd, B.A., Maxwell, E.S., Leschziner, A.E., Unger, V.M. and Baserga, S.J. (2009) A dimeric structure for archaeal box C/D small ribonucleoproteins. Science **325**, 1384–1387
106. Lin, J., Lai, S., Jia, R., Xu, A., Zhang, L., Lu, J. and Ye, K. (2011) Structural basis for site-specific ribose methylation by box C/D RNA protein complexes. Nature **469**, 559–563

107. Bower-Phipps, K. and Taylor, D. (2012) The box C/D sRNP dimeric architecture is conserved across domain Archaea. RNA **18**, 1527–1540
108. Cahill, N.M., Friend, K., Speckmann, W., Li, Z.H., Terns, R.M., Terns, M.P. and Steitz, J.A. (2002) Site-specific cross-linking analyses reveal an asymmetric protein distribution for a box C/D snoRNP. EMBO J. **21**, 3816–3828
109. Qu, G., van Nues, R.W., Watkins, N.J. and Maxwell, E.S. (2011) The spatial-functional coupling of box C/D and C′/D′ RNPs is an evolutionarily conserved feature of the eukaryotic box C/D snoRNP nucleotide modification complex. Mol. Cell. Biol. **31**, 365–374
110. Li, S., Duan, J., Li, D. and Yang, B. (2011) Reconstitution and structural analysis of the yeast box H/ACA RNA-guided pseudouridine synthase. Genes Dev. **25**, 2409–2421
111. Walbott, H., Machado-Pinilla, R., Liger, D., Blaud, M., Réty, S., Grozdanov, P.N., Godin, K., van Tilbeurgh, H., Varani, G., Meier, U.T. and Leulliot, N. (2011) The H/ACA RNP assembly factor SHQ1 functions as an RNA mimic. GenesDev. **25**, 2398–2408
112. Baird, N.J., Zhang, J., Hamma, T. and Ferré-D'Amaré, A.R. (2012) YbxF and YlxQ are bacterial homologs of L7Ae and bind K-turns but not K-loops. RNA **18**, 759–770
113. Boulon, S., Marmier-Gourrier, N., Pradet-Balade, B., Wurth, L., Verheggen, C., Jády, B.E., Rothé, B., Pescia, C., Robert, M.C., Kiss, T. et al. (2008) The Hsp90 chaperone controls the biogenesis of L7Ae RNPs through conserved machinery. J. Cell Biol. **180**, 579–595
114. Cho, I.-M., Lai, L.B., Susanti, D., Mukhopadhyay, B., and Gopalan, V. (2010) Ribosomal protein L7Ae is a subunit of archaeal RNase P. Proc. Natl. Acad. Sci. U.S.A. **107**, 14573–14578
115. Starostina, N.G., Marshburn, S., Johnson, L.S., Eddy, S.R., Terns, R.M. and Terns, M.P. (2004) Circular box C/D RNAs in *Pyrococcus furiosus*. Proc. Natl. Acad. Sci. U.S.A. **101**, 14097–14101
116. Danan, M., Schwartz, S., Edelheit, S. and Sorek, R. (2012) Transcriptome-wide discovery of circular RNAs in Archaea. Nucleic Acids Res. **40**, 3131–3142
117. Yang, C.-Y., Zhou, H., Luo, J. and Qu, L.-H. (2005) Identification of 20 snoRNA-like RNAs from the primitive eukaryote, *Giardia lamblia*. Biochem. Biophys. Res. Commun. **328**, 1224–1231
118. Liang, X.-H., Uliel, S., Hury, A., Barth, S., Doniger, T., Unger, R. and Michaeli, S. (2005) A genome-wide analysis of C/D and H/ACA-like small nucleolar RNAs in *Trypanosoma brucei* reveals a trypanosome-specific pattern of rRNA modification. RNA **11**, 619–645
119. Nakaar, V., Dare, A. and Hong, D. (1994) Upstream tRNA genes are essential for expression of small nuclear and cytoplasmic RNA genes in trypanosomes. Mol. Cell. Biol. **14**, 6736–6742
120. Hertel, J., Hofacker, I.L. and Stadler, P.F. (2008) SnoReport: computational identification of snoRNAs with unknown targets. Bioinformatics **24**, 158–164
121. Schattner, P., Brooks, A.N. and Lowe, T.M. (2005) The tRNAscan-SE, snoscan and snoGPS web servers for the detection of tRNAs and snoRNAs. Nucleic Acids Res. **33**, W686–W689
122. Yang, J.-H., Zhang, X.-C., Huang, Z.-P., Zhou, H., Huang, M.B., Zhang, S., Chen, Y.Q. and Qu, L.H. (2006) snoSeeker: an advanced computational package for screening of guide and orphan snoRNA genes in the human genome. Nucleic Acids Res. **34**, 5112–5123
123. Muller, S., Charpentier, B., Branlant, C. and Leclerc, F. (2007) A dedicated computational approach for the identification of archaeal H/ACA sRNAs. In RNA Modification, (Gott, J., ed.), pp. 355–387, Academic Press
124. Tafer, H., Kehr, S., Hertel, J., Hofacker, I.L. and Stadler, P.F. (2010) RNAsnoop: efficient target prediction for H/ACA snoRNAs. Bioinformatics **26**, 610–616
125. Kehr, S., Bartschat, S., Stadler, P.F. and Tafer, H. (2011) PLEXY: efficient target prediction for box C/D snoRNAs. Bioinformatics **27**, 279–280
126. Myslyuk, I., Doniger, T., Horesh, Y., Hury, A., Hoffer, R., Ziporen, Y., Michaeli, S. and Unger, R. (2008) Psiscan: a computational approach to identify H/ACA-like and AGA-like non-coding RNA in trypanosomatid genomes. BMC Bioinf. **9**, 471
127. Brown, J.W., Clark, G.P., Leader, D.J., Simpson, C.G., and Lowe, T. (2001) Multiple snoRNA gene clusters from *Arabidopsis*. RNA **7**, 1817–1832

128. Piekna-Przybylska, D., Decatur, W.A. and Fournier, M.J. (2007) New bioinformatic tools for analysis of nucleotide modifications in eukaryotic rRNA. RNA **13**, 305–312
129. Bleichert, F. and Baserga, S. (2011) Small ribonucleoproteins in ribosome biogenesis. In Protein Reviews: the Nucleolus, (Olson, M.O.J., ed.), pp. 135–156, Springer New York, New York
130. Watkins, N.J., Dickmanns, A. and Lu, R. (2002) Conserved stem II of the box C/D motif is essential for nucleolar localization and is required, along with the 15.5 K protein, for the hierarchical assembly of the box C/D snoRNP. Mol. Cell. Biol. **22**, 8342–8352

Role of small nuclear RNAs in eukaryotic gene expression

Saba Valadkhan[1] and Lalith S. Gunawardane

Center for RNA Molecular Biology, Case Western Reserve University, 10900 Euclid Avenue, Cleveland, OH 44106, U.S.A.

Abstract

Eukaryotic cells contain small, highly abundant, nuclear-localized non-coding RNAs [snRNAs (small nuclear RNAs)] which play important roles in splicing of introns from primary genomic transcripts. Through a combination of RNA–RNA and RNA–protein interactions, two of the snRNPs, U1 and U2, recognize the splice sites and the branch site of introns. A complex remodelling of RNA–RNA and protein-based interactions follows, resulting in the assembly of catalytically competent spliceosomes, in which the snRNAs and their bound proteins play central roles. This process involves formation of extensive base-pairing interactions between U2 and U6, U6 and the 5′ splice site, and U5 and the exonic sequences immediately adjacent to the 5′ and 3′ splice sites. Thus RNA–RNA interactions involving U2, U5 and U6 help position the reacting groups of the first and second steps of splicing. In addition, U6 is also thought to participate in formation of the spliceosomal active site. Furthermore, emerging evidence suggests additional roles for snRNAs in regulation of various aspects of RNA biogenesis, from transcription to polyadenylation and RNA stability. These snRNP-mediated regulatory roles probably serve to ensure the co-ordination of the different processes involved in biogenesis of RNAs and point to the central importance of snRNAs in eukaryotic gene expression.

Keywords:
group II intron, ribozyme, small nuclear RNA, spliceosome.

[1]*To whom correspondence should be addressed (email saba.valadkhan@case.edu).*

Introduction: the challenge of splicing and evolution of eukaryotic snRNAs (small nuclear RNAs)

A fascinating feature of modern eukaryotic genes is the nearly ubiquitous presence of intervening sequences or introns, which interrupt the continuity of the information content of genes. Thus, before primary gene transcripts can be used by the cell, introns must be accurately removed or 'spliced'[1]. In addition, recent research indicates that introns themselves often harbour regulatory or otherwise functional sequences, and their accurate and timely removal is often critical for their cellular function [2]. The intronic sequences in higher eukaryotic genes are much longer than the non-intronic sequences, the exons, and the sequence-based information that specifies the intron–exon boundaries is highly complex and poorly understood. Thus accurately distinguishing these two sets of functional sequences that co-exist in eukaryotic primary transcripts is a highly challenging task for the eukaryotic gene expression machinery.

Although modern mammals have one of the most complex splicing patterns among extant eukaryotes, it is likely that even in primordial eukaryotes splicing was already a highly complex process. On the basis of currently accepted models of evolution of eukaryotes, introns probably originated from self-splicing ribozymes that dated from pre-cellular life and constituted the majority of the genomes of ancient eukaryotes [3]. Later on, probably in order to prevent genomic instability, the introns lost their self-splicing capacity and, instead, the splicing function was delegated to a cellular machine, the spliceosome, which acted *in trans* to remove introns from primary transcripts. Although the origin and evolution of the early spliceosomes is still largely mysterious, several lines of evidence suggest that they probably evolved from self-splicing introns [4]. This hypothesis is partly based on the fact that the mechanism of intron removal by the spliceosome, performed through two consecutive transesterification reactions resulting in removal of a branched lariat intron, is identical with the splicing reaction performed by a class of extant self-splicing introns called the group II introns [5] (Figures 1 and 2). These introns, which are found in all three kingdoms of life, are RNA-centric catalytic sequences composed of a number of base-paired RNA structures called 'domains' (Figure 2) [6]. Extensive research has elucidated the identity of the catalytically essential sequences in these introns (Figure 2) [6]. Intriguingly, the RNA components of the spliceosome, the snRNAs, show unmistakable similarities to fragments of the catalytically essential domains of group II introns in sequence, secondary structure and function. Of the five major spliceosomal snRNAs (U1, U2, U4, U5 and U6), three of them (U2, U5 and U6) have clear structural and functional similarities to critical domains of group II introns (Figure 2). Domain-swapping experiments have indicated that isolated domains of group II introns and U5 and U6 snRNA substructures could functionally replace each other, proving their functional equivalence [5,6]. Another set of sequences in group II introns are functionally equivalent to U1 snRNA, despite the lack of structural similarity (Figure 2). Currently there are no known functional or structural equivalents for the U4 snRNA in group II introns and the evolutionary origin of this snRNA is completely unknown. On the other hand, a number of domains of the group II introns do not have an equivalent among the spliceosomal snRNAs and it is likely that, in the spliceosome, these RNA domains are replaced by spliceosomal proteins. While the possibility

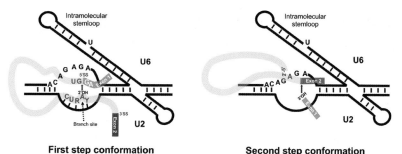

Figure 1. U6 and U2 snRNAs and the mRNA at the time of first and second steps of splicing

The location of U6, U2 and the U6 ISL is shown. The intron is shown by a thick light blue line connecting the two exons. Position of the 5′ splice site (5′SS), 3′ splice site (3′SS) and branch site are shown. Solid arrows point to the site of the nucleophilic attack during the two steps of splicing. The first step involves a nucleophilic attack by the 2′ hydroxy group of a specific adenosine residue in the intron, the branch site adenosine (the bulged A), on the 5′ splice site. This leads to a trans-esterification reaction in which the 2′ oxygen of the branch-site adenosine replaces the 3′ oxygen of the last nucleotide of the upstream exon. The result of this reaction is the release of the first exon and the formation of an unusual 2′–5′ linkage between the branch site adenosine and the first nucleotide of the intron (right-hand panel). During the second step, the free 3′ hydroxyl moiety of the newly released exon is activated for a similar nucleophilic attack on the 3′ splice site, resulting in ligation of the two exons and release of the intron as a branched lariat. Base-pairing interactions are shown by short black lines. The location of the 2′–5′ linkage formed after the first step of splicing at the branch site is shown.

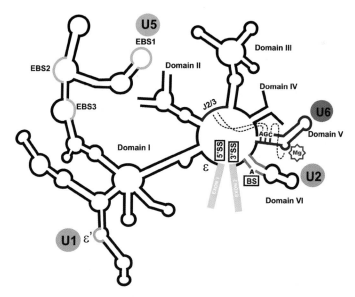

Figure 2. The structural organization of the group II self-splicing intron aI5γ

The location of domains I–VI, the two exons (shown in green), the splice sites and the branch site (5′SS, 3′SS and BS respectively) are shown. The position of J2/3 and the AGC sequence are indicated. The metal-binding site of domain V is shown by a red 'Mg' sign. Broken lines connect regions which are juxtaposed to form the catalytic core. The circles denote the functional equivalent of each domain or subdomain in the spliceosome. ε and ε′ sites, which are involved in an interaction important in recognition of the 5′ splice site, are shown.

© 2013 Biochemical Society

of convergent evolution cannot be formally ruled out, the above-mentioned similarities strongly suggest that at least a number of snRNAs are evolutionary remnants of primordial self-splicing ribozymes.

Roles of snRNAs in the spliceosome

The snRNAs were first discovered in 1970s as small highly abundant nuclear RNAs which formed the core of RNP (ribonucleoprotein) particles which showed strong reactivity with the immune sera from patients with autoimmune disorders [7,8]. Further analysis indicated the presence of sequence complementarity between one of the snRNAs, U1, and the sequences found at the 5′ splice site of primary transcripts, ultimately leading to the discovery of their involvement in splicing [9]. Further research identified a second set of snRNAs, named U11, U12, U4atac and U6atac, which are functional counterparts of U1, U2, U4 and U6 snRNAs respectively, and participate in the formation of a 'minor' spliceosome which is responsible for removal of an atypical subset of introns, most of which have alternative consensus sequences at the splice sites and branch site [10]. U5 snRNA is found in both 'major' and 'minor' spliceosomes. Each of the nine spliceosomal snRNAs are stably associated with a set of proteins, creating the snRNP (small nuclear ribonucleoprotein) particles which form the main functional subunits of the spliceosome. Analysis of the spliceosomal function suggests that snRNPs and several non-snRNP spliceosomal proteins assemble on each intron in a stepwise elaborate fashion through a large number of conformational rearrangements which start from the recognition of the splice junctions and culminate in splicing catalysis, followed by disassembly and recycling of the spliceosomal components [1,11] (Figure 3).

Recognition of the 5′ splice site by U1 snRNP

Recognition of introns in primary transcripts is partly mediated by detection of 'consensus' sequences found at the junction of introns and exons, the 5′ and 3′ splice sites. These consensus sequences are rather short: 5′-AG/GURAGU and 5′-YAG/G for 5′ and 3′ splice sites respectively, where R denotes either a G or an A, and Y denotes a C or a U and / marks the location of the splice site [1,11]. Interestingly, in higher eukaryotes and especially in mammals, the splice site sequences are highly degenerate and, in many cases, significantly deviate from the consensus sequence, thus necessitating an elaborate multi-step recognition mechanism mediated by a combination of RNA–RNA base pairing and RNA–protein interactions [1,11].

The association of U1 snRNP and its functional equivalent in the 'minor' spliceosomes, U11, with the 5′ splice site is one of the earliest and arguably most important events in the spliceosomal assembly pathway [11,12] (Figure 3, the E complex). The recognition and binding of the 5′ splice site is mediated both by base pairing of a single-stranded sequence at the 5′-end of U1 snRNA to the 5′ splice site and through an intricate web of interactions between the pre-mRNA and U1C, a U1-specific protein [5,11,13]. Interestingly, a high–resolution structure of U1 snRNP indicated the presence of a number of interactions between U1C and the nucleotides at the 5′ end of U1 which base pair to the 5′ splice site, thus providing a structural basis for the dual RNA–protein recognition of the 5′ splice site [12,13]. While this RNA–protein recognition of the 5′ splice site by U1 is functionally critical for the majority of cellular transcripts *in vivo*, several other proteins also contribute to the selection of the 5′ splice site [12]. Furthermore, at least some primary transcripts can be spliced in the absence of U1 snRNP *in*

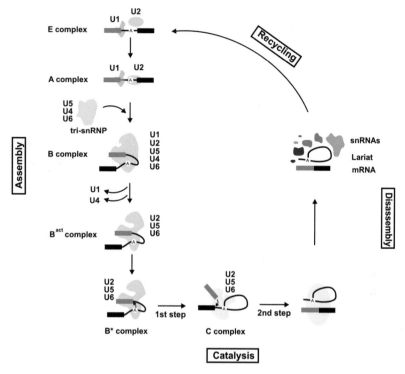

Figure 3. The spliceosomal cycle
The spliceosomal complexes formed during a splicing cycle are shown. The snRNAs present in each splicing complex are indicated. The exons on the pre-mRNA are shown as blue and black rectangles, with the intron drawn as a thin line connecting the two. The position of the branch-site adenosine is marked.

vitro, pointing to extensive redundancy in the splicing machinery. Finally, the binding of U1 is not necessarily synonymous with productive splicing, since the binding of U1 to sequences involved in negative regulation of splicing has been documented [12]. Current data suggest that the binding of U1 to such elements is important in splicing regulation and exclusion of pseudo splice sites, underscoring the importance of sequence context in splicing.

The U2 snRNP and recognition of the branch site and 3′ splice site

Once bound to the 5′ splice site in a sequence context which is conducive to splicing, U1 helps initiate the spliceosomal assembly by forming a network of interactions with U2 snRNP that plays a dominant role in recognition of the 3′ splice site and the branch site, another region in the introns which is recognized by the spliceosomes [12,14]. Branch sites, which are typically located ~30 nt from the 3′-end of introns, contain the adenosine which acts as the nucleophile of the first transesterification step of splicing and forms the branched lariat structure found in the splicing intermediates and post-splicing introns (Figures 1 and 3). In early spliceosomes, U2 snRNP is loosely associated with the end of the intron through protein-mediated interactions. However, in an ATP-dependent step which involves the displacement of intron-bound proteins and remodelling of base-pairing interactions within the U2 snRNA, U2 forms a stable interaction with the branch site and 3′ splice site [11,13,15] (Figure 3, the A complex). This

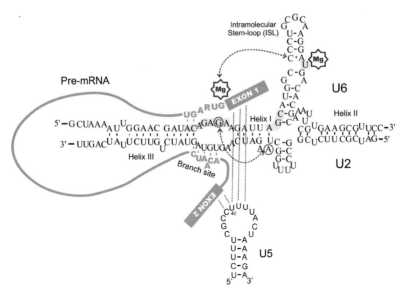

Figure 4. The known RNA–RNA interactions at the spliceosomal catalytic core at the time of the first step of splicing
The central domain of human U6 and the 5′ domain of human U2, which contain the sequences necessary for splicing *in vivo*, are shown. The base-pairing interactions between U2 and U6 which form helices I, II and III are shown. The exons in pre-mRNA are drawn as rectangles, with the intron as a thin solid line. The sequence of the branch site of the intron is shown. The conserved stem loop I of U5 snRNA is shown at the bottom, with the thin broken lines marking non-canonical base pairs between the exonic sequences next to the splice sites and the U5 loop I. The ACAGAGA and AGC domains are highlighted in yellow. The dotted curved red line joining circled nucleotides points to a tertiary interaction detected in activated spliceosomes. The other dotted red line connects the binding sites for two functionally required metal ions (shown by red 'Mg' signs) which may be located near each other in activated spliceosomes.

interaction is partly mediated through a base-pairing interaction between U2 and sequences flanking the branch site adenosine (Figure 4), and is stabilized by several RNA–protein interactions. The base pairing between U2 and the branch site leaves the branch site adenosine in an unpaired extrahelical conformation necessary for efficient splicing (Figure 4) [5,6]. In addition, the interaction between U1 and U2 snRNPs generates a loop that brings the 5′ and 3′ splice sites together and helps to 'define' the introns and exons [14].

Formation of a catalytically active spliceosome

Three of the snRNPs, U4, U5 and U6, form a ternary complex termed the 'tri-snRNP' and collectively integrate into the assembling spliceosomes [1,5] (Figure 3, the B complex). Of all of the spliceosomal snRNAs, U6 is the most conserved and contains two invariant domains, the ACAGAGA and AGC boxes, which play a critical functional role in splicing (Figure 4) [5,16]. Furthermore, it contains an ISL (intramolecular stem-loop) which is almost identical with the catalytic domain V of group II introns and, similar to its group II intron counterpart, binds a functionally required divalent cation, pointing to a critical role in splicing catalysis for this snRNA (Figures 2, 4 and 5) [5,16]. Perhaps in order to prevent it from prematurely forming a catalytically active structure, within the tri–snRNP U6 is kept in an inactive conformation

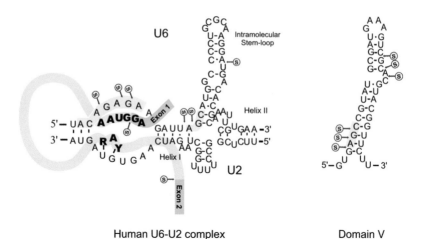

Figure 5. Structural and functional similarities between the catalytically crucial domain V of group II introns and U6 snRNA

The central domains of human U6 and the 5' domain of human U2 snRNAs are shown. The pre-mRNA is shown as a lighter green line, with the sequence of the branch site indicated. The ACAGAGA and AGC sequences in U6 and domain V are highlighted in yellow. The sites of phosphorothioate interference, which may point to metal-binding sites, are shown with red marks.

through a base-pairing interaction with U4 snRNA that prevents the formation of its functionally critical ISL. Current data do not indicate any additional functions for U4 snRNA except acting as a negative chaperon for U6. Once the base paired U4/U6 complex joins the spliceosome in association with U5 within the tri-snRNP, the U4/U6 duplex is unwound in a tightly controlled manner and U2 snRNA replaces U4 as the base-pairing partner of U6 (Figure 4) [1,5,16]. In addition, U6 replaces U1 at the 5' splice site, forming canonical and non-canonical base-pairing interactions with this sequence (Figure 3, B^{act} complex and Figure 4). The base-pairing interactions between U6 and U2 allow the formation of the U6 ISL and, further, serve to juxtapose the branch site, which is bound to the branch-binding sequence in U2, and the 5' splice site, which is bound by U6 (Figure 4). At the same time, U5 snRNA forms non-canonical base-pairing interactions with the exon sequences immediately adjacent to the splice sites and participates in aligning the exons to ensure their optimal positioning for the second step of splicing (Figure 4) [1,5,16]. These base-pairing rearrangements are accompanied by an extensive rearrangement of protein–protein and protein–RNA interactions, culminating in the formation of catalytically competent spliceosomes (B* and C complexes, Figure 3). U1 and U4 are not stably associated with fully assembled spliceosomes, thus leaving U2, U5 and U6 snRNAs as the only spliceosomal RNA components required for catalysis (Figure 4) [5]. Interestingly, as mentioned above, these three snRNAs have clear structural and functional counterparts in self-splicing group II introns, raising the possibility that the catalytic core of the two splicing systems may be closely similar (Figure 2) [5,6,17–19].

Catalysis of splicing: the role of snRNAs

Mutagenesis studies have shown that, at least *in vitro*, the conserved loop of U5, which was previously shown to be the functionally important domain of the molecule (Figure 4), was in fact dispensable for splicing [1,5]. On the other hand, it has been shown that, under certain conditions, several positions within the branch binding sequence of U2 snRNA can be mistakenly recognized as the 5′ splice site [20]. These results imply that the branch binding sequence of U2, which is functionally the most critical region of this snRNA, is not essential for spliceosomal catalysis, at least under certain conditions. As the rest of U2 seems to mainly fulfill structural roles by forming base-pairing interactions with other U2 sequences or with U6 snRNA [5,15], these results suggest that U6 snRNA may be the only RNA that is absolutely crucial for splicing catalysis, at least under the conditions studied so far.

Several additional lines of evidence suggest that U6 may form part of the catalytic domain of the spliceosome [5,16]. U6 is the most conserved of all spliceosomal snRNAs and several point mutations in its two evolutionarily invariant sequences, the ACAGAGA and AGC boxes, lead to a block in splicing, pointing to a critical function for these two sequences (Figures 4 and 5). Cross-linking and mutational complementation analyses have indicated that the first step of splicing occurs in close proximity to the ACAGAGA box, suggesting that this sequence is in the immediate vicinity of or even forms part of the spliceosomal active site. Current data suggest that in group II introns, the active site is formed by juxtaposition of the AGC triad and the asymmetric internal bulge of domain V along with a short purine-rich sequence (J2/3, Figure 2) which is considered functionally equivalent to the ACAGAGA box in U6 [19]. Interestingly, U6 contains the equivalent of all of these sequences which form the active site in group II introns, and close similarities in phosphorothioate interference patterns between catalytic domain V of group II introns and U6 (Figure 5) suggest that they may be functionally related [5,17–19]. Furthermore, hydroxyl radical footprinting and *in vivo* mutagenesis studies have pointed to the proximity of the AGC triad, the ACAGAGA box and the area near the bulged residue in ISL in functional spliceosomes (Figure 4) [5], perhaps in an arrangement similar to or even identical with the one found in group II introns [5,17–19].

Interestingly, analyses on *in-vitro*-transcribed, protein-free U6 and U2 snRNAs indicate that they can efficiently form a base-paired complex *in vitro* which in many respects resembles the one formed in the activated spliceosomes (Figure 4) [5,21–23]. Furthermore, it has been shown that the *in-vitro*-assembled human U2–U6 complex can indeed catalyse a two-step splicing reaction which closely resembles the one catalysed by the self-splicing group II introns and the spliceosome [24,25]. On the basis of the data above, the snRNAs seem to be fully competent to form the majority, if not all, of the spliceosomal active site and to perform catalysis, similar to the self-splicing group II introns, albeit with much lower efficiency.

If we assume that the spliceosome is an RNA catalyst, the snRNAs are unusual ribozymes in many respects, perhaps most importantly they are unusually small compared with other natural ribozymes catalysing splicing reactions. The larger size of other natural splicing ribozymes is thought to allow them to fold into complex tertiary structures, which in turn enable them to create sophisticated active sites necessary for such complex reactions. It is conceivable that due to their short length, the U6 and U2 snRNAs at best form an inefficient splicing ribozyme, which requires other spliceosomal factors for stable positioning of the active-site elements and the reacting groups. Although the exact role played by the proteins in the spliceosomal catalytic core is mostly unknown, their possible roles could range from assisting the snRNAs in assuming their functional structure, assisting in or independently co-ordinating critical metal ions and participating in the positioning of the substrates, to independently forming part of the active site and even direct involvement in catalysis [26].

Beyond splicing: other biological roles of snRNAs

Although the spliceosomal snRNAs (and their minor spliceosomal counterparts U11, U12, U4atac and U6 atac) play major roles in spliceosomal function, data suggest additional roles in regulation of gene expression for the snRNP particles. The interaction of U1 and the 5′ splice site, in addition to its function in splice-site selection, also seems to play a role in stabilization of some messages [12]. Both U1 and U2 snRNPs have been implicated in transcriptional regulation through stimulation of the rate of formation of the first phosphodiester bond at transcription initiation and also through interaction with a component of the pre-initiation complex, TFIIH (transcription factor II H) respectively [27,28]. U1 snRNP seems to also regulate the efficiency of polyadenylation via the interaction of a U1-specific protein, U1A, with a component of the CPSF (cleavage and polyadenylation stimulating factor) [12]. In addition, it has been reported that binding of U1 to a 5′ splice site-like sequence in the 3′-UTR (untranslated region) of some mRNAs inhibits their polyadenylation, leading to degradation of the RNA [12]. Thus the snRNAs and their bound proteins seem to act in co-ordination of the various steps in gene expression, in addition to playing the central role in splicing.

Another spliceosomal snRNP, the SL (spliced leader) particle, plays a critical role as the splice donor in a non-canonical *trans*-splicing reaction mainly observed in some protozoa and lower invertebrates [29]. In the *trans*-splicing reaction, SL is treated as a mini exon plus a short intron, with the exonic sequences 'spliced' *in trans* to a 3′ splice site on the primary transcripts of *trans*-splicing organisms. Thus, unlike the other spliceosomal snRNAs, SL is consumed during the *trans*-splicing reaction.

Although the majority of snRNAs play a role in splicing, there are other abundant small nuclear-localized RNAs which play critical roles in other cellular processes. A non-spliceosomal snRNA, U7, functions in 3′-end processing of replication dependent histone mRNAs, which are not polyadenylated and instead terminate in a conserved stem-loop structure (SL element) generated by endonucleolytic cleavage of the pre-mRNA [30]. Similar to the spliceosomal snRNAs, U7 also forms an snRNP by associating with a set of proteins which together form the so-called Sm ring, and is recruited to histone pre-mRNA primarily through base-pairing interactions via its 5′-end with a purine-rich HDE (histone downstream element) which is located in the vicinity of the cleavage site. Together with a protein which binds the SL element, U7 recruits a complex that triggers endonucleolytic cleavage between SL and the HDE by the CPSF73 endonuclease, thus forming the mature histone mRNA.

Conclusion

As detailed above, the ability of snRNAs to form strong, specific interactions via base pairing with another RNA is extensively utilized in the spliceosome and during processing of the histone 3′-ends. Base-pairing interactions contribute to substrate recognition (U1, U2 and U7), positioning of the branch site in a strained catalytic bulged conformation (U2), regulation of the activity of another snRNA (U4) and juxtaposition of reactive substrates (U2, U5 and U6). Since RNA–RNA interactions similar to those formed by some of the spliceosomal snRNAs play identical or closely related roles in group II introns, it is conceivable that the robustness of RNA–RNA interactions has led to their preservation throughout the evolution of the spliceosome from group II-like ancient ribozymes. Although RNAs can perform the above-mentioned tasks with ease and even more effectively than proteins, when it comes to catalysis, proteins seem to have an advantage over RNA, at least in the case of natural ribozymes. The evolutionary reason behind the preservation of U6 snRNAs as a constituent of the spliceosomal catalytic core remains an open question.

Another feature of the snRNAs is their participation in multiple sets of base-pairing interactions that at times are mutually exclusive and, thus, act as switches between different functional states. The presence of such interactions underscores the highly complex evolutionary pressures under which the snRNAs have evolved. In addition, most of the snRNAs bind a number of proteins which play important functional roles. While in many cases these proteins complement the function of the snRNA to which they bind, emerging evidence suggests that they can impart a completely novel function on the snRNP. Research in the coming years is likely to provide additional instances of multi-functionality of the snRNPs and further elucidate their contribution to the highly complex network of interactions which regulate eukaryotic gene expression.

Summary

- The majority of cellular snRNAs (small nuclear RNAs) function in splicing, with their ability to form specific base-pairing interactions extensively utilized in recognition of functional sequence elements in primary transcripts. Examples include the interactions between U1 and U6 and the 5′ splice site, U2 and the branch site, and U5 and the exonic sequences.
- Another set of RNA–RNA interactions occur between snRNAs and plays important roles in regulation of the timing of their folding into their active structure (in the case of the interaction between U4 and U6) or form a structural scaffold for juxtaposition of the reactive groups in the splicing reaction (formation of the U2/U6 base-paired complex).
- In addition to forming RNA–RNA interactions, U6 snRNA seems to play a critical role in catalysis of the splicing reaction. Whether the spliceosomal proteins play a role in catalysis remains to be determined.
- Another aspect of the function of snRNAs is forming interactions with a set of proteins which play important roles in spliceosomal assembly, regulation and co-ordination of splicing and other steps of gene expression.

References

1. Will, C.L. and Luhrmann, R. (2006) Spliceosome structure and function. In The RNA World (Gesteland, R.F., Cech, T.R. and Atkins, J.F., eds), pp. 369–400, Cold Spring Harbor Laboratory Press
2. Chorev, M. and Carmel, L. (2012) The function of introns. Front. Genet. **3**, 55

3. Roy, S.W. and Irimia, M. (2009) Splicing in the eukaryotic ancestor: form, function and dysfunction. Trends Ecol. Evol. **24**, 447–455
4. Rodríguez-Trelles, F., Tarrío, R. and Ayala, F.J. (2006) Origins and evolution of spliceosomal introns. Annu. Rev. Genet. **40**, 47–76
5. Valadkhan, S. (2010) Role of the snRNAs in spliceosomal active site. RNA Biol. **7**, 345–353
6. Pyle, A.M. and Lambowitz, A.M. (2006) Group II introns: ribozymes that splice RNA and invade DNA. In The RNA World. (Gesteland, R.F., Cech, T.R. and Atkins, J.F., eds), pp. 469–506, Cold Spring Harbor Laboratory Press
7. Lerner, M.R. and Steitz, J.A. (1979) Antibodies to small nuclear RNAs complexed with proteins are produced by patients with systemic lupus erythematosus. Proc. Natl. Acad. Sci. U.S.A. **76**, 5495–5499
8. Zieve, G. and Penman, S. (1976) Small RNA species of the HeLa cell: metabolism and subcellular localization. Cell **8**, 19–31
9. Lerner, M.R., Boyle, J.A., Mount, S.M., Wolin, S.L. and Steitz, J.A. (1980) Are snRNPs involved in splicing? Nature **283**, 220–224
10. Will, C.L. and Lührmann, R. (2005) Splicing of a rare class of introns by the U12-dependent spliceosome. Biol. Chem. **386**, 713–724
11. Wahl, M.C., Will, C.L. and Lührmann, R. (2009) The spliceosome: design principles of a dynamic RNP machine. Cell **136**, 701–718
12. Buratti, E. and Baralle, D. (2010) Novel roles of U1 snRNP in alternative splicing regulation. RNA Biol. **7**, 412–419
13. Valadkhan, S. and Jaladat, Y. (2010) The spliceosomal proteome: at the heart of the largest cellular ribonucleoprotein machine. Proteomics **10**, 4128–4141
14. Shao, W., Kim, H.-S., Cao, Y., Xu, Y.-Z. and Query, C.C. (2012) A U1-U2 snRNP interaction network during intron definition. Mol. Cell. Biol. **32**, 470–478
15. Perriman, R. and Ares, Jr, M. (2010) Invariant U2 snRNA nucleotides form a stem loop to recognize the intron early in splicing. Mol. Cell **38**, 416–427
16. Valadkhan, S. (2007) The spliceosome: a ribozyme at heart? Biol. Chem. **388**, 693–697
17. Michel, F., Costa, M. and Westhof, E. (2009) The ribozyme core of group II introns: a structure in want of partners. Trends Biochem. Sci. **34**, 189–199.
18. Dayie, K.T. and Padgett, R.A. (2008) A glimpse into the active site of a group II intron and maybe the spliceosome, too. RNA **14**, 1697–1703
19. Keating, K.S., Toor, N., Perlman, P.S. and Pyle, A.M. (2010) A structural analysis of the group II intron active site and implications for the spliceosome. RNA **16**, 1–9
20. Smith, D.J., Query, C.C. and Konarska, M.M. (2007) *Trans*-splicing to spliceosomal U2 snRNA suggests disruption of branch site–U2 pairing during pre–mRNA splicing. Mol. Cell **26**, 883–890
21. Valadkhan, S. and Manley, J.L. (2000) A tertiary interaction detected in a human U2–U6 snRNA complex assembled **in vitro** resembles a genetically proven interaction in yeast. RNA **6**, 206–219
22. Butcher, S.E. (2011) The spliceosome and its metal ions. Met. Ions Life Sci. **9**, 235–251
23. Guo, Z., Karunatilaka, K.S. and Rueda, D. (2009) Single-molecule analysis of protein-free U2-U6 snRNAs. Nat. Struct. Mol. Biol. **16**, 1154–1159
24. Jaladat, Y., Zhang, B., Mohammadi, A. and Valadkhan, S. (2011) Splicing of an intervening sequence by protein–free human snRNAs. RNA Biol. **8**, 372–377
25. Valadkhan, S., Mohammadi, A., Jaladat, Y. and Geisler, S. (2009) Protein-free small nuclear RNAs catalyze a two–step splicing reaction. Proc. Natl. Acad. Sci. U.S.A. **106**, 11901–11906
26. Hsieh, J., Andrews, A.J. and Fierke, C.A. (2004) Roles of protein subunits in RNA-protein complexes: lessons from ribonuclease P. Biopolymers **73**, 79–89
27. McKay, S.L. and Johnson, T.L. (2011) An investigation of a role for U2 snRNP spliceosomal components in regulating transcription. PLoS ONE **6**, e16077

28. Kwek, K.Y., Murphy, S., Furger, A., Thomas, B., O'Gorman, W., Kimura, H., Proudfoot, N.J. and Akoulitchev, A. (2002) U1 snRNA associates with TFIIH and regulates transcriptional initiation. Nat. Struct. Mol. Biol. **9**, 800–805
29. Lasda, E.L. and Blumenthal, T. (2011) Trans-splicing. Wiley Interdisciplinary Reviews: RNA **2**, 417–434
30. Ideue, T., Adachi, S., Naganuma, T., Tanigawa, A., Natsume, T. and Hirose, T. (2012) U7 small nuclear ribonucleoprotein represses histone gene transcription in cell cycle-arrested cells. Proc. Natl. Acad. Sci. U.S.A. **109**, 5693–5698

The functions of natural antisense transcripts

Megan Wight and Andreas Werner[1]

Institute for Cell and Molecular Biosciences, Newcastle University, Framlington Place, Newcastle NE2 4HH, U.K.

Abstract

NATs (natural antisense transcripts) are widespread in eukaryotic genomes. Experimental evidence indicates that sense and antisense transcripts interact, suggesting a role for NATs in the regulation of gene expression. On the other hand, the transcription of a gene locus in both orientations and RNA hybrid formation can also lead to transcriptional interference, trigger an immune response or induce gene silencing. Tissue-specific expression of NATs and the compartmentalization of cells ensure that the regulatory impact of NATs prevails. Consequently, NATs are now acknowledged as important modulators of gene expression. New mechanisms of action and important biological roles of NATs keep emerging, making regulatory RNAs an exciting and quickly moving area of research.

Keywords:
natural antisense transcript, RNA masking, transcription interference.

Introduction

NATs (natural antisense transcripts) were first described in bacteria as early as in 1981 and found to control plasmid numbers [1]. Since then, bacterial regulatory RNAs have been extensively studied. In general, they interfere with translation initiation and both inhibitory and stimulatory effects can be observed. Inhibitory NATs block or melt hairpin structures in mRNAs that are required for ribosome binding. Activation is observed when NATs counteract inhibitory structures and enable efficient translation initiation or elongation. The significance of RNA-mediated control mechanisms in maintaining homoeostasis in bacteria is unclear. An

[1]*To whom correspondence should be addressed (email andreas.werner@ncl.ac.uk).*

attractive hypothesis suggests that NAT-mediated regulation is not essential in single cells, but enables decisions that control collective behaviour such as biofilm formation, motility or virulence [2]. Biological roles of NATs and the related enzymatic mechanisms differ fundamentally between bacteria and higher eukaryotic systems. The present chapter focuses on the latter, although excellent reviews cover bacterial regulatory RNAs in detail [3].

Between 1986 and 2002, sporadic NATs were also discovered in eukaryotes, including mice and humans. The significance of the serendipitous findings was unclear and NATs were clearly thought to be an exception rather than the rule [4]. This image changed drastically with the start of the genomic age at the beginning of this century. Large-scale sequencing approaches, tiling arrays and data mining projects all identified staggering numbers of NATs, particularly in the transcriptomes of higher eukaryotes. The most comprehensive study was performed in mice by the FANTOM Consortium and reported that up to 72% of transcriptional units are transcribed in both orientations [5]. The figures reported for humans are considerably lower (40%) depending on different cell types [6,7]. In this context it is important to note that somewhat arbitrary parameters set to distinguish real NATs from experimental noise can skew the perceived scale of NATs. However, there is a consensus that natural antisense transcription is a pervasive and highly relevant phenomenon.

NATs can be categorized according to their mode of action or structure. Generally, NATs can act in *cis* and *trans*. *Cis* describes a regulatory impact that affects predominantly the corresponding sense transcript. Alternatively, NATs can regulate transcripts from other genomic loci, thus they act in *trans*. *Trans*-regulation is largely observed with short RNAs such as miRNAs (microRNAs) and endogenous siRNAs (small interfering RNAs) as well as with NATs transcribed from pseudogenes, which may interfere with their highly similar parent gene transcript [8]. The present chapter focuses on *cis*-acting NATs as the other groups of ncRNAs (non-coding RNAs) are discussed in Chapters 2, 3 and 8. *Cis*-NATs are transcribed by RNA Pol II (RNA polymerase II) and show hallmarks of mRNA processing such as splicing, a poly-A tail and a cap structure. The direction of transcription can either be convergent or divergent, resulting in sense/antisense transcript pairs that overlap either at the 3′- or the 5′-end respectively (Figure 1). In rare cases, NATs can be fully embedded within the corresponding sense gene. These transcripts may or may not share complementarity with the fully processed sense transcript [9,10].

From a functional point of view, NATs can stimulate or reduce the expression of the sense transcript. The terms concordant and discordant regulation are used to describe stimulatory or inhibitory effects of NATs respectively (Figure 1) [11]. Although the mechanisms involved in both concordant and discordant regulation are under intense investigation, key questions remain. For example, the finding that NATs have very low expression levels begs the question: why are sense transcripts expressed in orders of magnitude greater than their regulatory antisense counterparts? The next section of the present chapter discusses the possibilities of how antisense transcripts regulate gene expression and also points to the limitations of our current understanding.

Two biological phenomena in humans and mice that have been established to depend on expression of specific NATs are parental imprinting and compensatory X chromosome inactivation in females. Interestingly, in these cases, antisense transcription from one allele is essential to silence the corresponding sense transcript in *cis* while the sense transcript on the other allele stays active. These are arguably the best studied examples of epigenetic gene silencing that involve a NAT. There are additional aspects to imprinting and X chromosome inactivation

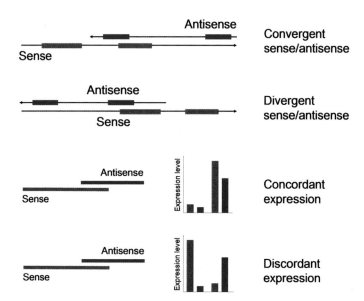

Figure 1. General genomic configuration of bi-directionally transcribed genes
Top two panels: sense is indicated in red, antisense in blue. Bottom panels: NATs can affect the expression of the corresponding sense transcript in a concordant (NAT ↑, sense transcript ↑) or discordant (NAT ↑, sense transcript ↓) manner.

such as allele choice or the spreading of the silencing mark to neighbouring genes that do not necessarily apply to the majority of NATs. For this reason it is unclear to what extent paradigms from imprinting and X chromosome inactivation translate to the general field of NATs. The present chapter focuses on this majority of NATs and we would like to refer the reader to excellent review articles for insights into imprinting and X chromosome dosage compensation [12].

How do NATs regulate gene expression?

Although bi-directionally transcribed genes are often manipulated experimentally to investigate sense–antisense interactions, the underlying mechanisms are still poorly understood. Research has implicated NATs in a range of mechanisms with various complexities, from the simple creation of a physical barrier against modifying factors to the direction of an intricate web of dynamic chromatin remodelling. Some of these processes are better established than others with underlying theories on the basis of a more sturdy foundation of evidence. It is important, however, to discuss them all to gauge the extent of this convoluted regulatory picture (Figure 2).

RNA masking

The first mechanism to be discussed centres on the formation of duplexes with sense and antisense pairs providing a physical barrier against post-transcriptional interactions. This 'masking' of a transcript blocks out factors that would otherwise induce splicing, influence stability or direct the RNA to specific cellular compartments (Figure 2) [9]. An increasing number of physiologically important sense–antisense pairs have recently been shown to act via RNA masking.

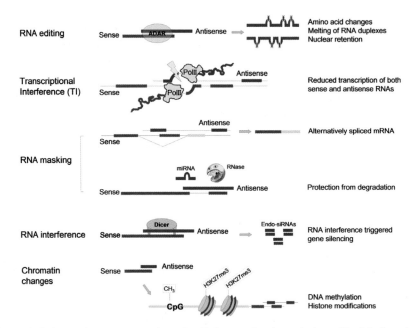

Figure 2. Schematic representation of cellular mechanisms induced by bi-directional transcription or NATs
The mechanisms include RNA editing, transcriptional interference, RNA masking, RNAi and chromatin modifications. Sense is indicated in red, antisense in blue. More details are given in the text.

Sense–antisense pairs tend to cross either one or more exon–intron border, the NAT therefore carries the potential to alter splicing of the sense transcript. This begs the question of whether there is an indirect relationship between alternative splicing and antisense regulation. Interesting as this theory may be, there are only two reported studies that demonstrate this link. The first example includes a NAT complementary to Zeb2 mRNA and its expression promotes the retention of a large intron at the 5′-end of the sense transcript. The intron contains an internal ribosome-binding site that enhances the expression of Zeb2. Being a transcriptional repressor, Zeb2 down-regulates E-cadherin (epithelial cadherin) expression, thereby inducing epithelial–mesenchymal transition [13].

The second example, TRα (thyroid hormone receptor α), expresses two isoforms, TRα1 and TRα2. The spliceform TRα1 is fully active, whereas the TRα2 only binds to the hormone without triggering a downstream response. The splicing of the primary transcript is regulated by a NAT RevErbAα whose expression correlates with the synthesis of the inactive receptor isoform. Interestingly, alternative splicing can also be induced by an oligonucleotide complementary to the crucial splice site [14].

These two examples seem to represent relatively isolated findings and little evidence so far suggests that NATs are heavily involved in the regulation of alternative splicing. Splice sites are not the only regulatory sites that can be affected by NATs; indeed, masking of regulatory sequences as well as interference with miRNA-binding sites by NATs has been reported in previous studies [15,16].

A case of RNA masking thought to have pathological significance is β-secretase mRNA. β-Secretase 1 is a key enzyme in the formation of amyloid protein fragments (Aβ1–40 and

Aβ1–42) and therefore closely linked to the pathophysiology of Alzheimer's disease. The expression of β-secretase is concordantly regulated by a processed and spliced antisense transcript. Neuroblastoma-derived cells (SH-SY5Y) show increased levels of both the antisense transcript and β-secretase in response to stressors known to promote Alzheimer's disease. Importantly, the antisense transcript was also found to be elevated in the brains of Alzheimer's patients [17]. A mechanism for how the NAT contributes to stabilization of the sense transcript was proposed by Faghihi et al. [16]. It is thought that the NAT competes with a repressive miRNA (*miR-485-5p*) for the same binding site, therefore stabilizing β-secretase mRNA [16]. miRNA–NAT competition could well be of general importance since a large proportion of NATs are complementary to the 3′-end of the corresponding sense transcript. On the other hand, the 3′-end of protein-coding transcripts also harbours most of the miRNA-binding sites.

Finally, a case of exposure rather than masking can be seen with aHIF (antisense hypoxia-inducible factor), a transcript complementary to HIF-1α mRNA. HIF-1α is a transcription factor induced by hypoxia and its expression is linked to tumour growth and progression. aHIF regulates HIF-1α in a discordant manner by exposing AU-rich sequences that destabilize the protein-coding sense transcript [15]. A general problem with RNA masking that is yet to be mentioned is in the formation of extended RNA duplexes. If found in the cytoplasm, these structures may be interpreted by the cell as signs of a viral infection and through PKR (double-stranded-RNA-dependent protein kinase) an immune response is triggered [18]. As a consequence, the cell containing the RNA hybrids will be destroyed, a result that cannot be the intention of a NAT-related 'regulatory' mechanism.

A-to-I editing

ADARs (adenosine deaminases that act on RNA) target dsRNAs (double-stranded RNAs). Any RNA that is double-stranded could be a potential substrate, revealing a mechanism for possible modifications of sense–antisense RNA duplexes. If such RNA pairs were to undergo A-to-I editing, it might follow that an antisense strand could manipulate the primary sequence of the sense partner as well as its localization or stability (Figure 2) [19,20].

Nevertheless, several studies have found that almost all A-to-I editing sites lie within inverted elements, mostly Alu repeats, in folded hairpin-like structures. Moreover, antisense transcripts showed few signs of editing sites other than in inverted repeats [21]. Despite these findings, it is important to acknowledge the chance of evidence being experimentally overlooked; edited sequences may swiftly degrade or remain in the nucleus and not be seen in expressed sequence datasets [22]. Nevertheless, it seems improbable that RNA editing plays a leading role in the regulatory functioning of NATs.

RNAi (RNA interference)

If sense and antisense RNAs are co-expressed in the same cell RNAi may be triggered. RNAi describes both an intrinsic method of post-transcriptional gene regulation and a process that is exploited experimentally to knock out genes of interest [23]. Central to this process are the type III ribonucleases Drosha and/or Dicer that cleave the dsRNA precursors (Figure 2). Whereas miRNA production involves a stem-loop RNA precursor and the action of Drosha and Dicer, the processing of sense–antisense RNA hybrids is likely to be performed by Dicer alone resulting in endo-siRNAs. The products are short oligonucleotides 21–23 nt in length. The short RNAs pair with a member of the Argonaute protein family to form the core of the RISC (RNA-induced silencing complex) [24].

In terms of the pairing with Argonautes, only one strand of the double-stranded short RNA precursor is selected as guide RNA, whereas the other so-called passenger strand unwinds and becomes degraded. The question of which strand becomes the guide and which takes the passenger role is particularly important in the context of endo-siRNAs. Strand selection dictates whether the sense or the antisense transcripts become the prime target for the newly formed RISC.

The involvement of RNAi in the biology of NATs is still contentious owing to the elusive nature of naturally produced endo-siRNAs. Only the latest transcriptome sequencing efforts are starting to find abundant endo-siRNAs in selected tissues, most prominently in testis [25,26]. Moreover, the simple expression of a naturally occurring sense–antisense pair from plasmids in a cell line does not necessarily result in the formation of siRNAs [27,28]. Therefore it could well be that NATs only trigger RNAi in selected cell types or at specific stages of development and differentiation [28].

TI (transcriptional interference)

Antisense transcripts do not necessarily require formation of a dsRNA duplex with their sense partner to exert a regulatory effect; the process of their transcription alone may be sufficient. There are a few mechanisms proposed for how transcription of one strand could suppress transcription of another in *cis*. These theories are collectively named TI [29]. TI in the initiation phase is proposed to involve the two promoters competing for use of regulatory elements and RNA Pol II. During the elongation phase, interference occurs in the form of a physical blockage: an oncoming RNA Pol complex from one strand halts progress of an RNA Pol II complex on the other stand (Figure 2). Alternatively, it may clear the promoter on the opposite strand [9]. Much of the complex web of regulatory mechanisms is still unclear. Since theories of TI are on the basis of limited experimental evidence, it is likely they only apply to a small number of NATs.

NAT-induced chromatin changes

Epigenetics describe the phenomenon of altered gene expression with inheritable consequences to the phenotype that do not involve changes to the DNA sequence. The best studied epigenetic marks are DNA methylation at cytosine (usually in a CG context) and histone modifications at various sites, notably methylation or acetylation of lysine residues. The modification of histones affect chromatin packaging and, depending on whether structures are tightened or loosened, repression or activation of gene expression are observed. Methylated DNA, on the other hand, interferes with the transcriptional machinery and also binds to repressor complexes containing enzymes that enable chromatin remodelling. Sequence specificity to these modifications is thought to be brought about by ncRNA. Accordingly, many of the mechanisms discussed above have been linked to epigenetic modifications at bi-directionally transcribed genomic loci. For example, hyper-edited RNA binds to a protein called vigilin, which in turn orchestrates a set of proteins that participate in chromatin silencing [30]. The link between antisense transcription and chromatin changes has been particularly well studied in parentally imprinted genes. Such genes are only expressed from one allele depending on the parental origin and characteristically express an antisense transcript. The importance of the antisense RNA was convincingly shown for the imprinted locus *Igf2r* (insulin-like growth factor 2 receptor)/*Airn*. A mouse was generated that expressed only a truncated short piece of the NAT by introducing an artificial polyadenylation site [31]. The transgenic animals show bi-allelic expression of the locus and reduced birth weight. The exact mechanism of how the antisense transcript *Airn* induces allelic silencing is still elusive. In this respect, another imprinted locus, *Kcnq1* and its antisense transcript *Kcnq1ot*, are better characterized: a model predicts that the antisense transcripts recruit a silencing complex that locally modifies chromatin. Interestingly, the process only works efficiently in the perinuclear space [32]. Epigenetic changes induced by NATs are not restricted to imprinted loci: for example in human leukaemic cells, the tumour suppressor gene p15 (*CDKN2B*; cyclin-dependent kinase 2 inhibitor B) has been found to be silenced by p15–NAT. A study using various reporter constructs demonstrated that p15–NAT expression led to heterochromatin formation and transcriptional silencing of the sense transcript [33]. Indeed, there is a rapidly growing list of specific NATs that repress the expression of the corresponding sense RNA at the transcriptional level. Silencing is brought on by repressive chromatin marks such as di- or tri-methylated histones (H3K9 and H3K27) or methylated DNA at CG residues (CpG methylation) in promoter regions (Figure 2) [34]. How exactly RNA recruits the different modifying complexes is not well understood, but it is an area of intense research.

NATs as drug targets

The link between NATs and epigenetic repression of protein-coding sense transcripts has prompted novel strategies to up-regulate clinically relevant genes. The idea is to knock down NATs to increase sense transcript levels and stimulate subsequent protein expression. The proof-of-concept for this approach has recently been delivered by Modarresi et al. [35] in a study focusing on BDNF (brain-derived neurotrophic factor). Knock down of the BDNF–NAT using siRNAs targeting areas outside the sense–antisense overlap led to a transient stimulation of both BDNF mRNA and protein. A similar effect was observed with RNA oligonucleotides containing locked nucleic acid modifications, so-called antagoNATs. They were demonstrated to be effective in both cell culture models and *in vivo*. AntagoNATs were administered directly into the brain of mice by a small peristaltic pump, resulting in increased BDNF expression and neuronal outgrowth [35]. A comparable approach was also used to stimulate the expression of β-secretase [17]. Aside from the considerable difficulties surrounding the delivery of active agents to the correct cells, antagoNATs may prove a novel and versatile weapon against a broad range of diseases.

Is there a bigger picture?

The number of regulatory NATs known to have physiological or pathophysiological significance is ever increasing. There is also substantial evidence to indicate a role of NATs in the evolution of highly complex organisms. The existence of a bigger picture is underpinned by the observation that NATs are selectively under-represented on mammalian X chromosomes

[10,36]. The X chromosome is only present in one active copy in mammalian cells: in females, the other X copy is epigenetically silenced and males carry a Y chromosome. The underrepresentation of NATs on the X chromosome may therefore reflect a strategy to avoid antisense-induced gene silencing. Autosomal genes, in contrast, are expressed from both alleles and aberrant expression of a NAT would only cause one allele to be silenced. The accuracy of this scenario is corroborated by the observations that first, such examples of monoallelic gene expression are quite frequent [37] and secondly, that these genes tend to express NATs [38].

In order to understand the bigger picture and to acknowledge the benefit of NAT-induced gene silencing, one has to focus on haploid developing sperm cells where NATs are

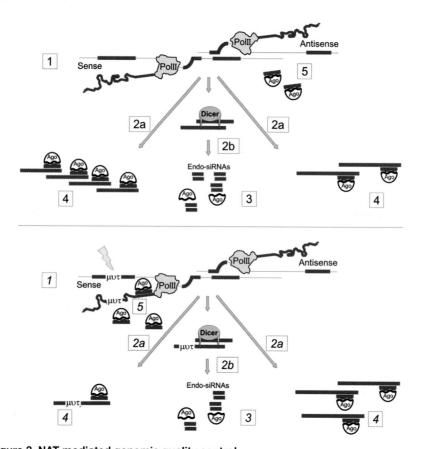

Figure 3. NAT-mediated genomic quality control
Top panel: the bi-directionally transcribed locus is intact. To run the control cycle, both sense and antisense transcripts are produced (1). Since the transcripts become fully processed, they can both persist as mRNAs (2a) or form RNA hybrids and feed into the RNAi pathway (2b). The resulting endo-siRNAs combine with Argonautes (most probably a non-slicing isoform such as Ago4) (3). According to the availability of target sequence, here the full-length sense and antisense transcripts, RISCs are formed (4). The spare Argonaute–endo-siRNA complexes (pre-RISC) will find the target sequence at the transcribed locus and induce transcriptional silencing (5). Bottom panel: the control cycle with a mutated sense transcript ($\mu\upsilon\tau$). The initial steps are identical with the top panel (1, 2 and 3); however, the mutation causes reduced levels of the full-length sense transcript (4). This could be the result of reduced transcription or transcript instability for example. As a consequence, pre-RISCs against the mutated transcript will reach the site of transcription and induce the silencing of the sense RNA (5) [39].

prominently expressed [38]. In a previously published theory [39], it was suggested that NATs play a role in the genomic quality control of developing sperm cells (Figure 3). According to this hypothesis, sense and antisense transcripts are co-expressed in haploid sperm cells and a proportion of the full-length transcripts is processed into endo-siRNAs. The ratio of protein-coding sense and non-coding antisense transcripts present is thought to determine strand selection of endo-siRNAs in the RISC. Accordingly, the RISC then targets the less abundant of the two transcripts which, in most cases, is the NAT. Consequently, an epigenetic signature is established that favours the expression of the protein-coding sense transcript and represses NAT. Any mutation that reduces the level of sense transcript might switch the balance in endo-siRNA strand selection and lead to the silencing of the protein-coding sense gene [39]. This will impinge on the development of the sperm cell and eventually positively select for cells without genomic damage.

Conclusions

NATs are arguably rising stars in the as yet rather dark universe of ncRNAs. This is an astonishing promotion, considering that only 8 years ago NATs were widely perceived as mere transcriptional noise. This dismissal was fuelled by the elusive nature of NATs. They are notoriously difficult to handle not only owing to their usually low expression level, but also because sense–antisense RNA pairs are tightly balanced and any experimental intervention distorts the system. Establishing physiological relevance of the observed effects often proves the greatest challenge. Nevertheless, a wealth of elegant studies have recently underpinned the biological importance of NATs and opened exciting and novel avenues for further research.

Summary
- Natural antisense transcripts constitute a distinct group within the ever increasing family of non-coding RNAs.
- Natural antisense transcripts influence the expression of the corresponding sense transcript.
- Several cellular mechanisms are triggered by natural antisense transcripts, most of them involving co-expression of sense–antisense transcripts and RNA hybrid formation.
- RNA–RNA and/or RNA–DNA hybrids may lead to RNA editing, RNA masking, RNA interference and eventually chromatin modifications.
- Strategies to knock down the natural antisense transipt with concomitant stimulation of protein encoding sense transcripts show great potential for medical applications.
- The abundance of natural antisense transcripts in the genomes of complex organisms may be related to a control mechanism that assesses the integrity of the coding transcriptome during sperm development.

References

1. Lacatena, R.M. and Cesareni, G. (1981) Base pairing of RNA I with its complementary sequence in the primer precursor inhibits ColE1 replication. Nature **294**, 623–626
2. Fender, A., Elf, J., Hampel, K., Zimmermann, B. and Wagner, E.G. (2010) RNAs actively cycle on the Sm-like protein Hfq. Genes Dev. **24**, 2621–2626
3. Storz, G., Vogel, J. and Wassarman, K.M. (2011) Regulation by small RNAs in bacteria: expanding frontiers. Mol. Cell **43**, 880–891
4. Vanhee-Brossollet, C. and Vaquero, C. (1998) Do natural antisense transcripts make sense in eukaryotes? Gene **211**, 1–9
5. Katayama, S., Tomaru, Y., Kasukawa, T., Waki, K., Nakanishi, M., Nakamura, M., Nishida, H., Yap, C.C., Suzuki, M., Kawai, J. et al. (2005) Antisense transcription in the mammalian transcriptome. Science **309**, 1564–1566
6. Engstrom, P.G., Suzuki, H., Ninomiya, N., Akalin, A., Sessa, L., Lavorgna, G., Brozzi, A., Luzi, L., Tan, S.L., Yang, L. et al. (2006) Complex loci in human and mouse genomes. PLoS Genet. **2**, e47
7. He, Y., Vogelstein, B., Velculescu, V.E., Papadopoulos, N. and Kinzler, K.W. (2008) The antisense transcriptomes of human cells. Science **322**, 1855–1857
8. Hawkins, P.G. and Morris, K.V. (2010) Transcriptional regulation of Oct4 by a long non-coding RNA antisense to Oct4-pseudogene 5. Transcription **1**, 165–175
9. Beiter, T., Reich, E., Williams, R.W. and Simon, P. (2008) Antisense transcription: a critical look in both directions. Cell. Mol. Life Sci. **66**, 94–112
10. Kiyosawa, H., Yamanaka, I., Osato, N., Kondo, S. and Hayashizaki, Y. (2003) Antisense transcripts with FANTOM2 clone set and their implications for gene regulation. Genome Res. **13**, 1324–1334
11. Faghihi, M.A. and Wahlestedt, C. (2009) Regulatory roles of natural antisense transcripts. Nat. Rev. Mol. Cell Biol. **10**, 637–643
12. Augui, S., Nora, E.P. and Heard, E. (2011) Regulation of X-chromosome inactivation by the X-inactivation centre. Nat. Rev. Genet. **12**, 429–442
13. Beltran, M., Puig, I., Pena, C., Garcia, J.M., Alvarez, A.B., Pena, R., Bonilla, F. and de Herreros, A.G. (2008) A natural antisense transcript regulates Zeb2/Sip1 gene expression during Snail1-induced epithelial-mesenchymal transition. Genes Dev. **22**, 756–769
14. Hastings, M.L., Ingle, H.A., Lazar, M.A. and Munroe, S.H. (2000) Post-transcriptional regulation of thyroid hormone receptor expression by cis-acting sequences and a naturally occurring antisense RNA. J. Biol. Chem. **275**, 11507–11513
15. Uchida, T., Rossignol, F., Matthay, M.A., Mounier, R., Couette, S., Clottes, E. and Clerici, C. (2004) Prolonged hypoxia differentially regulates hypoxia-inducible factor (HIF)-1α and HIF-2α expression in lung epithelial cells: implication of natural antisense HIF-1α. J. Biol. Chem. **279**, 14871–14878
16. Faghihi, M.A., Zhang, M., Huang, J., Modarresi, F., Van der Brug, M.P., Nalls, M.A., Cookson, M.R., St Laurent, 3rd, G. and Wahlestedt, C. (2010) Evidence for natural antisense transcript-mediated inhibition of microRNA function. Genome Biol. **11**, R56
17. Faghihi, M.A., Modarresi, F., Khalil, A.M., Wood, D.E., Sahagan, B.G., Morgan, T.E., Finch, C.E., St Laurent, 3rd, G., Kenny, P.J. and Wahlestedt, C. (2008) Expression of a noncoding RNA is elevated in Alzheimer's disease and drives rapid feed-forward regulation of β-secretase. Nat. Med. **14**, 723–730
18. Wang, Q. and Carmichael, G.G. (2004) Effects of length and location on the cellular response to double-stranded RNA. Microbiol. Mol. Biol. Rev. **68**, 432–452
19. Nishikura, K. (2010) Functions and regulation of RNA editing by ADAR deaminases. Annu. Rev. Biochem. **79**, 321–349
20. Zhang, Z. and Carmichael, G.G. (2001) The fate of dsRNA in the nucleus: a p54(nrb)-containing complex mediates the nuclear retention of promiscuously A-to-I edited RNAs. Cell **106**, 465–475

21. Levanon, E.Y., Eisenberg, E., Yelin, R., Nemzer, S., Hallegger, M., Shemesh, R., Fligelman, Z.Y., Shoshan, A., Pollock, S.R., Sztybel, D. et al. (2004) Systematic identification of abundant A-to-I editing sites in the human transcriptome. Nat. Biotechnol. **22**, 1001–1005
22. Neeman, Y., Dahary, D., Levanon, E.Y., Sorek, R. and Eisenberg, E. (2005) Is there any sense in antisense editing? Trends Genet. **21**, 544–547
23. Hannon, G.J. (2002) RNA interference. Nature **418**, 244–251
24. Czech, B. and Hannon, G.J. (2011) Small RNA sorting: matchmaking for Argonautes. Nat. Rev. Genet. **12**, 19–31
25. Tam, O.H., Aravin, A.A., Stein, P., Girard, A., Murchison, E.P., Cheloufi, S., Hodges, E., Anger, M., Sachidanandam, R., Schultz, R.M. et al. (2008) Pseudogene-derived small interfering RNAs regulate gene expression in mouse oocytes. Nature **453**, 534–538
26. Watanabe, T., Totoki, Y., Toyoda, A., Kaneda, M., Kuramochi-Miyagawa, S., Obata, Y., Chiba, H., Kohara, Y., Kono, T., Nakano, T. et al. (2008) Endogenous siRNAs from naturally formed dsRNAs regulate transcripts in mouse oocytes. Nature **453**, 539–543
27. Faghihi, M.A. and Wahlestedt, C. (2006) RNA interference is not involved in natural antisense mediated regulation of gene expression in mammals. Genome Biol. **7**, R38
28. Gullerova, M. and Proudfoot, N.J. (2012) Convergent transcription induces transcriptional gene silencing in fission yeast and mammalian cells. Nat. Struct. Mol. Biol. **19**, 1193–1201
29. Shearwin, K.E., Callen, B.P. and Egan, J.B. (2005) Transcriptional interference: a crash course. Trends Genet. **21**, 339–345
30. Wang, Q., Zhang, Z., Blackwell, K. and Carmichael, G.G. (2005) Vigilins bind to promiscuously A-to-I-edited RNAs and are involved in the formation of heterochromatin. Curr. Biol. **15**, 384–391
31. Sleutels, F., Zwart, R. and Barlow, D.P. (2002) The non-coding *Air* RNA is required for silencing autosomal imprinted genes. Nature **415**, 810–813
32. Kanduri, C. (2008) Functional insights into long antisense noncoding RNA Kcnq1ot1 mediated bidirectional silencing. RNA Biol. **5**, 208–211
33. Yu, W., Gius, D., Onyango, P., Muldoon-Jacobs, K., Karp, J., Feinberg, A.P. and Cui, H. (2008) Epigenetic silencing of tumour suppressor gene p15 by its antisense RNA. Nature **451**, 202–206
34. Malecova, B. and Morris, K.V. (2010) Transcriptional gene silencing through epigenetic changes mediated by non-coding RNAs. Curr. Opin. Mol. Ther. **12**, 214–222
35. Modarresi, F., Faghihi, M.A., Lopez-Toledano, M.A., Fatemi, R.P., Magistri, M., Brothers, S.P., van der Brug, M.P. and Wahlestedt, C. (2012) Inhibition of natural antisense transcripts *in vivo* results in gene-specific transcriptional upregulation. Nat. Biotechnol. **30**, 453–459
36. Chen, J., Sun, M., Kent, W.J., Huang, X., Xie, H., Wang, W., Zhou, G., Shi, R.Z. and Rowley, J.D. (2004) Over 20% of human transcripts might form sense–antisense pairs. Nucleic Acids Res. **32**, 4812–4820
37. Gimelbrant, A., Hutchinson, J.N., Thompson, B.R. and Chess, A. (2007) Widespread monoallelic expression on human autosomes. Science **318**, 1136–1140
38. Carlile, M., Swan, D., Jackson, K., Preston-Fayers, K., Ballester, B., Flicek, P. and Werner, A. (2009) Strand selective generation of endo-siRNAs from the Na/phosphate transporter gene *Slc34a1* in murine tissues. Nucleic Acids Res. **37**, 2274–2282
39. Werner, A. and Swan, D. (2010) What are natural antisense transcripts good for? Biochem. Soc. Trans. **38**, 1144–1149

Pseudogenes as regulators of biological function

Ryan C. Pink and David R.F. Carter[1]

School of Life Sciences, Oxford Brookes University, Gipsy Lane, Headington, Oxford OX3 0BP, U.K.

Abstract

A pseudogene arises when a gene loses the ability to produce a protein, which can be due to mutation or inaccurate duplication. Previous dogma has dictated that because the pseudogene no longer produces a protein it becomes functionless and evolutionarily inert, being neither conserved nor removed. However, recent evidence has forced a re-evaluation of this view. Some pseudogenes, although not translated into protein, are at least transcribed into RNA. In some cases, these pseudogene transcripts are capable of influencing the activity of other genes that code for proteins, thereby altering expression and in turn affecting the phenotype of the organism. In the present chapter, we will define pseudogenes, describe the evidence that they are transcribed into non-coding RNAs and outline the mechanisms by which they are able to influence the machinery of the eukaryotic cell.

Keywords:
non-coding RNA, pseudogene, RNA, transcription.

Introduction

A pseudogene is generally defined as a copy of a gene that has lost the capacity to produce a functional protein. They were first discovered in the 1970s when a copy of the 5S rRNA gene was found in *Xenopus laevis* with homology to the active gene, but with a clear truncation that rendered it non-functional [1]. Sporadic discovery and characterization of pseudogenes over the following 20 years has revealed a number of mechanisms for pseudogene formation [2]. Unitary pseudogenes are formed when spontaneous mutations occur in a coding gene that

---

[1]To whom correspondence should be addressed (email dcarter@brookes.ac.uk).

abolish either transcription or translation (Figure 1A) [3]. A second class of pseudogene, the duplicated pseudogene (Figure 1B), is formed when replication of the chromosome is performed incorrectly [2]. Such duplication events often lead to the formation of functional gene families, such as those found in the Hox gene clusters, but if part of the gene is not faithfully copied then these can lead to frameshift mutations or the loss of a promoter or enhancer, thus resulting in a non-functional duplicated pseudogene. The final class, known as the processed pseudogene (Figure 1C), is formed when an mRNA molecule is reverse-transcribed and integrated into a new location in the parental genome [4]. Because processed pseudogenes are produced from mRNA, they usually lack introns and a promoter, and are therefore only transcribed if they become integrated close to a pre-existing promoter [5].

The sequencing of a range of genomes, including the human genome, has revealed the extent of pseudogene abundance [6–8]. Estimates for the number of human pseudogenes range from 10000 to 20000, making them almost as prevalent as coding genes [9]. The majority of these are processed pseudogenes [6–8] and fewer than 100 are unitary pseudogenes [3]. Interestingly, the processed pseudogenes found in the human genome have been formed from just 10% of the coding genes [6,8], suggesting that either not all genes are capable of producing processed pseudogenes, or that only the processed pseudogenes produced by certain types of gene are selected for by evolution. The types of genes that produce processed pseudogenes are predominantly highly expressed housekeeping genes or shorter RNAs such as genes encoding ribosomal proteins [10]. It is of note that whereas mammalian genomes are particularly well endowed with pseudogene numbers [9], they are by no means the only species that harbour them. Pseudogenes have been found in various species [11], including bacteria, plants, insects and nematode worms, examples of which can be found in various databases [12].

Pseudogenes have often been labelled as 'junk DNA' because they lack protein-coding capacity. In fact, some genes that appear to be pseudogenized may in fact code for proteins

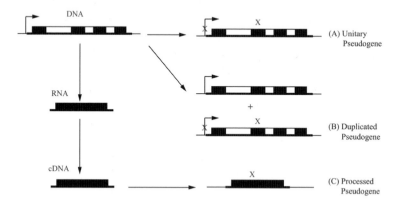

Figure 1. Different classes of pseudogene
(**A**) A unitary pseudogene is formed when a spontaneous mutation occurs in a coding gene. Such mutations may ablate transcription from the promoter or cause premature stop codons or frameshifts to occur. (**B**) A duplicated pseudogene is formed when a gene is duplicated, but in such a way that mutations in the copy prevent formation of a protein. (**C**) Processed pseudogenes arise when DNA is transcribed into RNA, which is then reverse-transcribed into copy DNA (cDNA) and integrated into the genome. Such pseudogenes often lack promoter activity and may have deletions or truncations that prevent protein formation. Closed boxes depict exons; open boxes depict introns; 'X' shows a mutation that prevents the DNA from being able to make a protein.

© The Authors Journal compilation © 2013 Biochemical Society

[11]. Others are genuinely non-coding, but are by no means 'junk', as they may actually play functional roles [11,13,14]. In particular, it is apparent that some pseudogenes are capable of producing lncRNAs [long ncRNAs (non-coding RNAs)] [11]. Although RNA was seen previously as taking a purely intermediary role in the expression of proteins from DNA, it is now widely acknowledged that ncRNAs can play significant roles in the regulation of gene expression [15]. The present chapter explores the evidence that some pseudogenes can regulate gene expression via the generation of ncRNAs.

Transcription

Examples of functional ncRNAs are being discovered at an ever-increasing rate, and their mechanism of function is diverse. Although the DNA encoding pseudogenes can play a role in normal biology, for example in the generation of antibody diversity [13], it is likely that the majority of pseudogene function is mediated through RNA molecules. Analysing the prevalence of pseudogene transcription should therefore give us insight into their potential function.

The way in which pseudogenes are generated often leads to their transcriptional silencing. Duplicated pseudogenes may be formed with mutations or deletions in their promoters, and processed pseudogenes may become integrated into transcriptionally silent regions of the genome. Assessing levels of pseudogene RNA can be difficult due to close homology with the original gene from which it was copied. Nevertheless, it has been demonstrated that many pseudogenes are transcribed into RNA, including the pseudogene versions of the housekeeping gene *GAPDH* (glyceraldehyde-3-phosphate dehydrogenase) [16], the transcription factor *Oct4* [17] and the tumour suppressor *PTEN* (phosphatase and tensin homologue deleted on chromosome 10) [18]. Microarray technology and next-generation DNA sequencing experiments now give us tools to analyse pseudogene transcription in a genome-wide manner; results using such techniques suggest that as many as one-fifth of pseudogenes may be transcribed into RNA [5]. Recent RNA-seq experiments have shown that pseudogene RNA represents a significant proportion of the transcriptome in cancer cells [19]. A recent genome-wide study of pseudogene sequences revealed that some are relatively well conserved, and these are more likely to be transcribed [20]. Furthermore, approximately half of transcribed pseudogenes identified in humans are well conserved across primates [21].

Analysing the expression of coding genes in different tissues, dynamically during development or specific biological responses, or during disease, can give insight into their function. The same principle also holds true for pseudogenes and other genes encoding ncRNAs [11]. It is worth noting that the high contribution of pseudogene RNAs to the transcriptional landscape of the cell makes designing primers for PCR (for analysing the activity of genes or pseudogenes) particularly challenging. Several pseudogenes exhibit tissue-specific patterns of transcription, with a particular prevalence for testis-specific expression [5,22]. Pseudogenes can also exhibit patterns of expression that are distinct from the parent coding genes from which they were copied [23,24]. Pseudogene RNA levels can also change during differentiation [25] and in diseases such as cancer [26] and diabetes [27]. The findings that pseudogene transcription can be conserved across millions of years, and can occur in a dynamic and tissue-specific manner suggests that the transcripts generated from pseudogenes may have an important role.

Evidence for function

To clearly demonstrate a genuine biological role for any gene it is not sufficient to show a correlation. In the last 15 years a number of functional experiments have been carried out that support a biological role for pseudogene RNA molecules in the regulation of their protein-coding counterparts [11,13]. Several genes associated with cancer progression have pseudogenes that may contribute to the pathophysiology of tumours. The ABC (ATP-binding cassette) transporters are a family of transmembrane channels involved in the transport of various solutes across membranes and whose deregulation is associated with drug resistance in tumours. *ABCC6* is one member of this family and, thanks to a well-conserved promoter, shares a similar pattern of tissue-specific expression to a pseudogene (*ABCC6P1*). The mRNA levels of the *ABCC6* gene are reduced when transcript levels of the *ABCC6* pseudogene are specifically reduced [28]. *Oct4* is a pluripotency-associated transcription factor that is involved in stem cell identity and is often deregulated during cancer progression. Overexpression of an *Oct4* pseudogene (*Oct4P1*) increased stem cell proliferation and inhibited differentiation of the mesenchymal lineage [29]. Overexpressing the pseudogene of the proto-oncogene *BRAF* led to increased MAPK (mitogen-activated protein kinase) signalling and a transformed phenotype in the mouse cell line NIH 3T3, and also caused tumour formation in mice [30]. In these cases the mechanisms by which the pseudogenes regulate the coding genes is unclear, but in other experiments a range of mechanisms are emerging that pseudogenes and their transcripts use to regulate gene expression and in turn cellular processes.

Antisense pairing and siRNA (small interfering RNA) production

If a processed pseudogene is integrated close to a promoter then this can result in transcription of the pseudogene. If the processed pseudogene is integrated in reverse orientation relative to the promoter then this would lead to an antisense transcript of the pseudogene, which, if it retains significant homology with the parent gene, could hybridize with the sense mRNA from the original coding gene [31]. Such an interaction occurs between the mRNA for the *nNOS* (neural nitric oxide synthase) gene and the RNA produced from a related pseudogene, which is transcribed in the antisense direction [32]. When both are transcribed in the same neurons of the snail *Lymnaea stagnalis*, the two form a duplex which leads to reduced translation of the coding gene (Figure 2A) [32]. Similarly when a transcript that is produced in the antisense direction to an *Oct4* pseudogene is blocked this leads to reduced levels of *Oct4* expression [33].

The pairing of sense and antisense transcripts leads to the formation of dsRNA (double-stranded RNA), which can trigger activation of the RNAi (RNA interference) pathway. In mouse oocytes it has been shown that pairing of antisense pseudogene RNA and sense coding-gene mRNA leads to the formation of such duplexes [34,35]. Dicer, a protein component of the RNAi pathway, slices the dsRNA into smaller fragments known as siRNAs. These siRNAs are incorporated into the RISC (RNA-induced silencing complex) and lead to the degradation of mRNA from the parental coding gene. For example, siRNAs were produced when mRNA from the *Ppp4r1* (encoding a protein phosphatase) gene and an antisense RNA from a pseudogene with high homology were combined; these siRNAs then appear to repress *Ppp4r1* expression

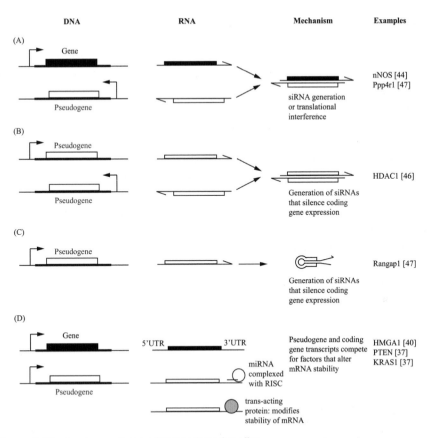

Figure 2. Mechanisms of pseudogene functionality
(**A**) Pseudogene RNA transcribed in the reverse (antisense) direction can combine with forward (sense) transcripts from the coding gene to produce dsRNA. This can inhibit translation of the coding RNA, or produce siRNAs that go into the RNAi pathway and cause the coding RNA to be degraded. siRNAs that destroy the coding transcript can also be generated by (**B**) pairing between sense and antisense transcribed pseudogenes and (**C**) double-stranded regions formed by secondary structure within a single pseudogene transcript. (**D**) Pseudogene transcripts may share binding sites for miRNAs or *trans*-acting proteins that regulate the stability of the mRNA. Increased levels of pseudogene transcripts can compete for these factors and therefore shield the coding transcripts from their effects.

(Figure 2A) [35]. Interestingly, the siRNAs generated did not always come from the pairing of a pseudogene RNA with a coding gene mRNA. Sometimes they were generated from the pairing of two pseudogenes (one transcribed in the sense direction and the other in the antisense), but the siRNA then represses the coding parent gene, such as in the case of *HDAC1* (encoding a histone deacetylase enzyme) (Figure 2B) [34]. In other instances the siRNAs were generated from the internal pairing of different regions within the same pseudogene transcript (i.e. from double-stranded regions formed by secondary structure folding). An example of the latter is the formation of hairpin loop structures in the *Au76* pseudogene RNA, which are processed into siRNAs that repress expression of the homologous coding gene *Rangap1* (encoding a protein that regulates G-coupled receptor signalling) (Figure 2C) [35]. Other organisms, including rice [36] and trypanosomes [37] have been shown to generate siRNAs from pseudogenes, which have the potential to repress expression of the parent coding gene, suggesting that this mechanism of pseudogene function may be relatively widespread in nature.

Regulation of mRNA stability

The regulation of mRNA stability is one way in which gene expression can be controlled. The stability of an mRNA can be influenced by protein factors that bind at different locations in the RNA [38]. If a pseudogene has a high homology with the parent coding gene, including the presence of the same *cis*-elements, then the RNAs from both could compete for the same pool of *trans*-acting molecules. Increasing the transcription of the pseudogene could produce a 'sink' for these *trans*-acting molecules, effectively lowering the concentration of the free proteins and thus changing the stability of mRNA from the coding genes (Figure 2D). This mechanism has been suggested for regulation of the imprinted *Makorin-1* gene (encoding an enzyme that adds an ubiquitin moiety to other proteins) by the related pseudogene *Makorin1-p1* [39]. Deregulation of the chromatin-related protein HMGA1 (high-mobility group A1) is involved in the development of Type 2 diabetes mellitus. In two diabetes patients, a low level of HMGA1 was found in correlation with a high level of the HMGA1 pseudogene RNA [27]. Blocking the pseudogene RNA partially restored the level of HMGA1 protein, suggesting that both transcripts compete for a positively stabilizing protein factor; when the level of pseudogene RNA increases it sequesters the protein factor, thus lowering the concentration of the free protein and causing a destabilization of *HMGA1* mRNA [27]. The *MYLKP1* (myosin light chain kinase pseudogene) gene is transcribed at higher levels in cancer cells. Overexpression of the pseudogene leads to destabilization of the parental gene mRNA and an increase in proliferation [40].

Another class of molecules that affects mRNA stability is the miRNA (microRNA). miRNAs are small (21–22 nt) single-stranded ncRNA molecules that are incorporated into the RISC and repress the expression of specific genes. This repression is achieved by base pairing between regions of the miRNA and the mRNA leading to degradation of the mRNA [41]. A single miRNA can target hundreds of different genes and any given mRNA can be targeted by more than one miRNA. Just as pseudogene RNAs can act as a 'sink' to sequester proteins that regulate coding mRNA stability, so they can also act as decoys to draw miRNAs away from coding genes (Figure 2D). A striking example of this was demonstrated for the *PTEN* gene and a related pseudogene *PTENP1* [26]. *PTEN* is a tumour suppressor gene whose expression level must be carefully regulated; even minor reductions in PTEN protein abundance can influence cancer initiation and severity [42]. The level of homology between *PTEN* and *PTENP1* is highest at the 3′-UTR (untranslated region), which is of significance because most miRNA–mRNA interactions are thought to occur at the 3′-UTR of the mRNA. Reducing the levels of *PTENP1* RNA leads to lower levels of *PTEN* mRNA and protein, and an inhibition of cell growth [26]. Overexpression of the 3′-UTR of *PTENP1* led to the reverse effect, with increased levels of *PTEN* expression and a stimulation of cell division [26]. These results suggest that specific miRNAs bind to both the *PTEN* and *PTENP1* 3′-UTR regions, and that increasing the amount of 'decoy' pseudogene transcript causes the miRNAs to be sequestered, lowering the effective concentration of the free miRNA in the cell and thus lifting the repression on the coding gene mRNA. This is consistent with the finding that *PTEN* and *PTENP1* expression levels are usually correlated in prostate cancer samples and that *PTENP1* is often deleted in sporadic colon cases [26]. A similar mechanism-of-action has been suggested for other gene–pseudogene pairs, including the oncogene *KRAS1* and the homologous pseudogene *KRASP1* [26]. Pseudogenes (or any ncRNAs) that act in this way to sequester miRNAs have been described as 'ceRNAs' (competing endogenous RNAs) [43]. The implications of these findings extend beyond the activity of individual pseudogenes, suggesting that the regulation of any given gene is partially dependent on the complex interactions of many RNA molecules (coding and non-coding) throughout the genome.

Conclusion

Genome sequencing has revealed an apparent paradox: the genomes of higher organisms such as humans do not have significantly more genes than lower organisms such as the nematode worm. To reconcile this it has been suggested that in higher organisms the greater abundance of regulatory ncRNAs allows the cell to fine-tune the expression of genes more precisely, thus orchestrating a more complex phenotype from the same number of building blocks [44]. Some unicellular organisms do not tolerate the formation of pseudogenes and instead actively remove them [45]. Mammals, and primates in particular, seem to retain and to some extent conserve pseudogenized genes [11,21]. This, coupled with the finding that many pseudogenes are transcribed, is consistent with pseudogenes playing a role as ncRNAs in regulating the activity of coding genes. It is unlikely that all pseudogenes play functional roles, but it appears that higher organisms have mechanisms in place that are ready to harness and conserve pseudogenes when one spontaneously arises that confers a useful regulatory role. Functional experiments have revealed that some pseudogenes do indeed play biological roles in cells, using a variety of mechanisms to influence genes and therefore affecting the phenotype of various organisms. With further experiments using next-generation sequencing technologies, the true extent of pseudogene influence and the mechanisms they use should be revealed.

Summary
- Pseudogenes are copies of genes that have lost the ability to produce a functional protein.
- Because they do not produce a protein, pseudogenes are often thought of as evolutionary relics, but evidence is emerging that some can play functional roles.
- Many pseudogenes are transcribed into RNA, and it is already known that some non-coding RNAs play a role in regulating gene expression.
- Many pseudogene RNAs appear capable of repressing or activating protein-coding genes with which they share sequence homology.
- The mechanisms of pseudogene function are varied, but often involve regulating the stability of the coding gene mRNA.
- This can be achieved by the pseudogene-mediated generation of small interfering RNAs, which knock down the coding gene via the RNA interference pathway, or by pseudogene transcript-mediated depletion of protein factors or microRNAs that affect coding mRNA stability.

We thank members of the laboratory for the critical reading of the chapter before submission and apologise to those whose excellent work has not been described due to space constraints. This work was supported by grants from Sparks and the Cancer and Polio Research Fund.

References

1. Jacq, C., Miller, J. and Brownlee, G. (1977) A pseudogene structure in 5S DNA of *Xenopus laevis*. Cell **12**, 109–120
2. Mighell, A.J., Smith, N.R., Robinson, P.A. and Markham, A.F. (2000) Vertebrate pseudogenes. FEBS Lett. **468**, 109–114
3. Zhang, Z.D., Frankish, A., Hunt, T., Harrow, J. and Gerstein, M. (2010) Identification and analysis of unitary pseudogenes: historic and contemporary gene losses in humans and other primates. Genome Biol. **11**, R26
4. D'Errico, I., Gadaleta, G. and Saccone, C. (2004) Pseudogenes in metazoa: origin and features. Briefings Funct. Genomics Proteomics **3**, 157–167
5. Zheng, D., Frankish, A., Baertsch, R., Kapranov, P., Reymond, A., Choo, S.W., Lu, Y., Denoeud, F., Antonarakis, S.E., Snyder, M. et al. (2007) Pseudogenes in the ENCODE regions: consensus annotation, analysis of transcription, and evolution. Genome Res. **17**, 839–851
6. Ohshima, K., Hattori, M., Yada, T., Gojobori, T., Sakaki, Y. and Okada, N. (2003) Whole-genome screening indicates a possible burst of formation of processed pseudogenes and Alu repeats by particular L1 subfamilies in ancestral primates. Genome Biol. **4**, R74
7. Torrents, D., Suyama, M., Zdobnov, E. and Bork, P. (2003) A genome-wide survey of human pseudogenes. Genome Res. **13**, 2559–2567
8. Zhang, Z., Harrison, P.M., Liu, Y. and Gerstein, M. (2003) Millions of years of evolution preserved: a comprehensive catalog of the processed pseudogenes in the human genome. Genome Res. **13**, 2541–2558
9. Zhang, Z. and Gerstein, M. (2004) Large-scale analysis of pseudogenes in the human genome. Curr. Opin. Genet. Dev. **14**, 328–335
10. Zhang, Z., Harrison, P. and Gerstein, M. (2002) Identification and analysis of over 2000 ribosomal protein pseudogenes in the human genome. Genome Res. **12**, 1466–1482
11. Pink, R.C., Wicks, K., Caley, D.P., Punch, E.K., Jacobs, L. and Carter, D.R. (2011) Pseudogenes: pseudo-functional or key regulators in health and disease? RNA **17**, 792–798
12. Karro, J.E., Yan, Y., Zheng, D., Zhang, Z., Carriero, N., Cayting, P., Harrrison, P. and Gerstein, M. (2007) Pseudogene.org: a comprehensive database and comparison platform for pseudogene annotation. Nucleic Acids Res. **35**, D55–D60
13. Balakirev, E.S. and Ayala, F.J. (2003) Pseudogenes: are they 'junk' or functional DNA? Annu. Rev. Genet. **37**, 123–151
14. Tutar, Y. (2012) Pseudogenes. Comp. Funct. Genomics **2012**, 424526
15. Caley, D., Pink, R., Trujillano, D. and Carter, D. (2010) Long noncoding RNAs, chromatin, and development. Sci. World J. **10**, 90–102
16. Tso, J.Y., Sun, X.H., Kao, T.H., Reece, K.S. and Wu, R. (1985) Isolation and characterization of rat and human glyceraldehyde-3-phosphate dehydrogenase cDNAs: genomic complexity and molecular evolution of the gene. Nucleic Acids Res. **13**, 2485–2502
17. Redshaw, Z. and Strain, A.J. (2010) Human haematopoietic stem cells express Oct4 pseudogenes and lack the ability to initiate Oct4 promoter-driven gene expression. J. Negat. Results Biomed. **9**, 2
18. Fujii, G.H., Morimoto, A.M., Berson, A.E. and Bolen, J.B. (1999) Transcriptional analysis of the PTEN/MMAC1 pseudogene, psiPTEN. Oncogene **18**, 1765–1769
19. Kalyana-Sundaram, S., Kumar–Sinha, C., Shankar, S., Robinson, D.R., Wu, Y.M., Cao, X., Asangani, I.A., Kothari, V., Prensner, J.R., Lonigro, R.J. et al. (2012) Expressed pseudogenes in the transcriptional landscape of human cancers. Cell **149**, 1622–1634

20. Pei, B., Sisu, C., Frankish, A., Howald, C., Habegger, L., Mu, X.J., Harte, R., Balasubramanian, S., Tanzer, A., Diekhans, M. et al. (2012) The GENCODE pseudogene resource. Genome Biol. **13**, R51
21. Khachane, A.N. and Harrison, P.M. (2009) Assessing the genomic evidence for conserved transcribed pseudogenes under selection. BMC Genomics **10**, 435
22. Reymond, A., Marigo, V., Yaylaoglu, M.B., Leoni, A., Ucla, C., Scamuffa, N., Caccioppoli, C., Dermitzakis, E.T., Lyle, R., Banfi, S. et al. (2002) Human chromosome 21 gene expression atlas in the mouse. Nature **420**, 582–586
23. Elliman, S.J., Wu, I. and Kemp, D.M. (2006) Adult tissue-specific expression of a Dppa3-derived retrogene represents a postnatal transcript of pluripotent cell origin. J. Biol. Chem. **281**, 16–19
24. Olsen, M.A. and Schechter, L.E. (1999) Cloning, mRNA localization and evolutionary conservation of a human 5-HT7 receptor pseudogene. Gene **227**, 63–69
25. Lin, M., Pedrosa, E., Shah, A., Hrabovsky, A., Maqbool, S., Zheng, D. and Lachman, H.M. (2011) RNA-seq of human neurons derived from iPS cells reveals candidate long non-coding RNAs involved in neurogenesis and neuropsychiatric disorders. PLoS ONE **6**, e23356
26. Poliseno, L., Salmena, L., Zhang, J., Carver, B., Haveman, W.J. and Pandolfi, P.P. (2010) A coding-independent function of gene and pseudogene mRNAs regulates tumour biology. Nature **465**, 1033–1038
27. Chiefari, E., Iiritano, S., Paonessa, F., Le Pera, I., Arcidiacono, B., Filocamo, M., Foti, D., Liebhaber, S.A. and Brunetti, A. (2010) Pseudogene-mediated posttranscriptional silencing of HMGA1 can result in insulin resistance and Type 2 diabetes. Nat. Commun. **1**, 1–7
28. Piehler, A.P., Hellum, M., Wenzel, J.J., Kaminski, E., Haug, K.B., Kierulf, P. and Kaminski, W.E. (2008) The human ABC transporter pseudogene family: evidence for transcription and gene–pseudogene interference. BMC Genomics **9**, 165
29. Lin, H., Shabbir, A., Molnar, M. and Lee, T. (2007) Stem cell regulatory function mediated by expression of a novel mouse Oct4 pseudogene. Biochem. Biophys. Res. Commun. **355**, 111–116
30. Zou, M., Baitei, E.Y., Alzahrani, A.S., Al-Mohanna, F., Farid, N.R., Meyer, B. and Shi, Y. (2009) Oncogenic activation of MAP kinase by BRAF pseudogene in thyroid tumors. Neoplasia **11**, 57–65
31. McCarrey, J.R. and Riggs, A.D. (1986) Determinator–inhibitor pairs as a mechanism for threshold setting in development: a possible function for pseudogenes. Proc. Natl. Acad. Sci. U.S.A. **83**, 679–683
32. Korneev, S.A., Park, J.H. and O'Shea, M. (1999) Neuronal expression of neural nitric oxide synthase (nNOS) protein is suppressed by an antisense RNA transcribed from an NOS pseudogene. J. Neurosci. **19**, 7711–7720
33. Hawkins, P.G. and Morris, K.V. (2010) Transcriptional regulation of Oct4 by a long non-coding RNA antisense to Oct4-pseudogene 5. Transcription **1**, 165–175
34. Tam, O.H., Aravin, A.A., Stein, P., Girard, A., Murchison, E.P., Cheloufi, S., Hodges, E., Anger, M., Sachidanandam, R., Schultz, R.M. and Hannon, G.J. (2008) Pseudogene-derived small interfering RNAs regulate gene expression in mouse oocytes. Nature **453**, 534–538
35. Watanabe, T., Totoki, Y., Toyoda, A., Kaneda, M., Kuramochi-Miyagawa, S., Obata, Y., Chiba, H., Kohara, Y., Kono, T., Nakano, T. et al. (2008) Endogenous siRNAs from naturally formed dsRNAs regulate transcripts in mouse oocytes. Nature **453**, 539–543
36. Guo, X., Zhang, Z., Gerstein, M.B. and Zheng, D. (2009) Small RNAs originated from pseudogenes: cis- or trans-acting? PLoS Comput. Biol. **5**, e1000449
37. Wen, Y.Z., Zheng, L.L., Liao, J.Y., Wang, M.H., Wei, Y., Guo, X.M., Qu, L.H., Ayala, F.J. and Lun, Z.R. (2011) Pseudogene-derived small interference RNAs regulate gene expression in African *Trypanosoma brucei*. Proc. Natl. Acad. Sci. U.S.A. **108**, 8345–8350
38. Ross, J. (1996) Control of messenger RNA stability in higher eukaryotes. Trends Genet. **12**, 171–175

39. Hirotsune, S., Yoshida, N., Chen, A., Garrett, L., Sugiyama, F., Takahashi, S., Yagami, K., Wynshaw-Boris, A. and Yoshiki, A. (2003) An expressed pseudogene regulates the messenger RNA stability of its homologous coding gene. Nature **423**, 91–96
40. Han, Y.J., Ma, S.F., Yourek, G., Park, Y.D. and Garcia, J.G. (2011) A transcribed pseudogene of MYLK promotes cell proliferation. FASEB J. **25**, 2305–2312
41. Bartel, D.P. (2009) microRNAs: target recognition and regulatory functions. Cell **136**, 215–233
42. Alimonti, A., Carracedo, A., Clohessy, J.G., Trotman, L.C., Nardella, C., Egia, A., Salmena, L., Sampieri, K., Haveman, W.J., Brog,i, E. et al. (2010) Subtle variations in Pten dose determine cancer susceptibility. Nat. Genet. **42**, 454–458
43. Salmena, L., Poliseno, L., Tay, Y., Kats, L. and Pandolfi, P.P. (2011) A ceRNA hypothesis: the Rosetta stone of a hidden RNA language? Cell **146**, 353–358
44. Mattick, J. (2007) A new paradigm for developmental biology. J. Exp. Biol. **210**, 1526–1547
45. Kuo, C.H. and Ochman, H. (2010) The extinction dynamics of bacterial pseudogenes. PLoS Genet. **6**, e1001050

Identification and function of long non-coding RNAs

Robert S. Young*[1] and Chris P. Ponting†[1]

*MRC Human Genetics Unit, Western General Hospital, Crewe Road, Edinburgh EH4 2XU, U.K.
†MRC Functional Genomics Unit, Department of Physiology, Anatomy and Genetics, University of Oxford, Oxford OX1 3PT, U.K.

Abstract

It is now clear that eukaryotic cells produce many thousands of non-coding RNAs. The least well-studied of these are longer than 200 nt and are known as lncRNAs (long non-coding RNAs). These loci are of particular interest as their biological relevance remains uncertain. Sequencing projects have identified thousands of these loci in a variety of species, from flies to humans. Genome-wide scans for functionality, such as evolutionary and expression analyses, suggest that many of these molecules have functional roles to play in the cell. Nevertheless, only a handful of lncRNAs have been experimentally investigated, and most of these appear to possess roles in regulating gene expression at a variety of different levels. Several lncRNAs have also been implicated in cancer. This evidence suggests that lncRNAs represent a new class of non-coding gene whose importance should become clearer upon further experimental investigation.

Keywords:
cancer, cis-regulation, dosage compensation, evolutionary conservation, imprinting, intergenic, transcription.

Introduction

Transcription (so-called 'dark matter' [1]) outside the boundaries of currently annotated protein-coding gene exons has frequently been detected [2,3]. The biological significance of these transcripts, however, remains far from clear. Scientific opinions are divided between those

[1]Correspondence can be addressed to either author (email robert.young@igmm.ed.ac.uk or chris.ponting@dpag.ox.ac.uk).

advocating that all are functional to those proposing that, without strong experimental evidence, they should be considered as being functionally inert; others' opinions lie between these polar extremes [4–6]. Transcripts exceeding 200 nt in length and that are apparently non-coding may be categorized as lncRNAs [long ncRNAs (non-coding RNAs); also previously designated 'large' ncRNAs]. A subset of these which do not overlap known protein-coding gene loci are known as lincRNAs (long intergenic ncRNAs) [7]. These have been preferred for investigation because their transcripts and functions are more likely to be independent of known protein-coding genes. It is this subset of the larger class of lncRNAs that we shall mainly discuss in the present chapter. Two examples of mouse lncRNA loci are shown in Figure 1. Most such loci generate apparently non-coding transcripts and can be complex, with transcripts produced whose genomic sequences overlap on both strands. Non-coding transcript maps are also extensive: the majority of nucleotides have been suggested to be transcribed at some point during normal development in, for example, *Drosophila melanogaster* [8].

Metazoan genomes are currently predicted to contain thousands of these loci, from approximately 1119 in the fruitfly [9] to more than 8000 in the human genome [10,11]. Such loci can be described as genes since they show some of the transcriptional, chromatin and evolutionary features of protein-coding genes. Nevertheless, this should not be meant to imply that each is functional. For example, some transcripts (RNA molecules) may not themselves transact a function, even if the act of their transcription is functional, for example by transcriptional interference [7].

Like the majority of mRNAs, many lncRNAs are thought to be polyadenylated and transcribed by RNA Pol II (RNA polymerase II). Recent mouse and human lincRNA sets have been defined using chromatin immunoprecipitation experiments targeting H3K4me3 (histone H3 Lys^4 trimethylation) and H3K36me3 (histone H3 Lys^{36} trimethylation) modifications [12,13] which are markers for RNA Pol II activity. Such lncRNAs may be spliced, and show a tendency to be expressed in a low and tissue–specific manner, with many thought not to be

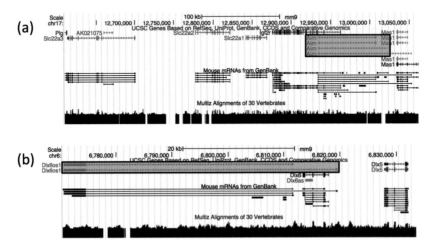

Figure 1. Two examples of mouse lncRNA loci whose transcripts' sequences overlap protein–coding loci
UCSC genes are shown in blue, with supporting mRNA sequence evidence in black and a conservation track across 30 vertebrate species is shown below. lncRNA loci are highlighted using yellow boxes. (**a**) *Airn*, an imprinted lncRNA locus. (**b**) *Evf2*, also known as *Dlx6os*. An antisense transcript *Dlx6as* is also apparent.

exported from the nucleus. Nevertheless, there is also evidence for abundant lncRNA transcripts that are not polyadenylated [14]. Little is yet known about such genes, and thus they are expected to be the subject of extensive study in the future years.

In the present chapter we start by describing how lncRNAs have been identified and computationally categorized, before then presenting evidence for molecular and cellular functions for a selection of lncRNAs and their possible involvement in disease processes.

lincRNA identification

Definition of intergenic transcripts

The first step in defining a set of lincRNAs is to identify transcripts that map to genomic regions lying outside the boundaries of currently annotated protein-coding gene models. The function of a mapped transcript that overlaps, on either strand, a protein-coding gene can be difficult to distinguish experimentally, using targeted knockout or knockdown approaches, from the function of this protein-coding gene. As a consequence, experimental investigation of lncRNAs has more often been focused on those that map to intergenic sequence. Determining whether a lncRNA locus is entirely intergenic is made more problematic because lncRNA transcripts are often incomplete and because they can emanate from a protein-coding gene's promoter or enhancer on either strand [15]. To increase the likelihood that the functions of a lncRNA locus are independent of those of its adjacent protein-coding genes, some studies have, at times, restricted their analyses to consider only lincRNA whose loci lie beyond a minimum distance from the nearest gene model in their analyses [16] or those whose orthologous sequences in related species are also non-protein-coding [12].

Intergenic transcripts can be detected experimentally using tiling microarrays [2]. Results from such experiments have been controversial [4], and early lincRNA collections thus relied primarily on sequenced cDNA and EST (expressed sequence tag) clones [17]. More recently, these catalogues have been superseded by lincRNAs derived from whole transcriptome sequencing (RNA-Seq). This approach generates millions of short (35–100 nt) sequence reads in parallel, and has confirmed that large amounts of intergenic sequence are transcribed into lincRNAs [11]. The high-throughput and relatively unbiased nature of this technique permits detailed assessment of the contribution of lincRNAs to the transcriptomes for a variety of tissues and/or species under different conditions.

Discrimination of coding from non-coding transcripts

Once a set of intergenic transcripts has been defined it is critical to separate those that have protein-coding potential from others that are true ncRNAs. It is relatively straightforward to assign transcripts with open reading frames exceeding 100 codons as being protein-coding [17]. Nevertheless, not all remaining transcripts will be non-coding, as they will also include transcripts encoding shorter polypeptides. To more accurately distinguish non-coding from coding transcripts, more sophisticated approaches have been developed. For example, the Coding Potential Calculator [18] considers six features of a transcript, including the proportion of the transcript covered by the candidate peptide-encoding region, and the polypeptide's sequence similarity to known proteins. An evolutionary approach, adopted by phyloCSF, predicts ncRNAs when their between-species sequence differences exhibit no bias as to whether they do or do not disrupt putatively encoded peptides [19].

Rather than relying on predictions, there will always be a preference for the protein-coding capability of a transcript to be determined experimentally. Large proteomic databases are now available for several species, and these can be used to investigate whether the RNA molecule is translated into protein. *In vitro* translation assays have been developed, but their results do not necessarily reflect *in vivo* biology. Associations between the candidate lincRNA and the ribosome can also be tested, with the expectation that a true lincRNA will not be translated and therefore would not be associated with this cellular organelle. Nevertheless, a study has reported that approximately half of a set of putative lincRNAs are ribosome-associated [20], leaving in doubt whether this test is accurate in discriminating coding from non-coding transcripts, or whether half of these transcripts are, instead, protein-coding. Experimental determination of an RNA sequence-dependent or -independent function for a transcript will be necessary for its assignment as a lincRNA. However, this may not always be sufficient because some transcripts will possess both RNA- and coding-sequence-dependent functions [21].

A computational or experimental method that discriminates accurately between coding and non-coding transcripts is thus currently lacking. A good compromise is to rely on *in silico* screens for protein-coding potential of putative lincRNAs, but to remain vigilant that these will contain false positive predictions, especially for genes encoding short polypeptides. A study of an individual lincRNA locus should seek to determine whether its mature transcribed RNA molecule is indeed the biologically relevant moiety or, instead, whether it is the act of transcription or the action of any short polypeptide encoded by the mRNA which is required for its function.

Genome-wide indicators of lincRNA functionality

Although many genomes that have been studied contain a considerable number of lincRNA loci, the proportion and number of these that are biologically functional have proved particularly controversial. Unlike for protein-coding genes, because the functional mechanisms of most non-coding transcripts or transcript regions are unknown, point mutation or deletion experiments are difficult to design, and their results are difficult to interpret. In addition, techniques such as RNAi (RNA interference), which reduce the abundance of a transcript, may result in no observable outcome when it is the act of transcription, rather than the mature transcript, that mediates functionality, or if only very low levels of the RNA transcript are required for functionality.

RNA structure

lincRNAs may fold into three-dimensional secondary structures that could be required for the mature RNA molecules to exert their functions. The ENCODE pilot study, which studied 1% of the human genome, predicted between 1500 and 1800 such structured regions [22]. However, there was little overlap in structure predictions produced using different methods, which suggests that they contain a large number of false-positive predictions.

Sequence conservation

An alternative, and complementary, approach to discriminate between functional and non-functional lincRNAs is to analyse their evolutionary signatures of functional constraint when comparing sequences among related species. This approach assumes that when a locus is functional then deleterious mutations in its sequence which disrupt this function will be preferentially purged from the population. Functional sequences therefore will show better conservation between species relative to neutrally evolving sequence. Some of the most well-studied lincRNAs (such as *Xist*, see below) are known to be poorly conserved between species. On a genomic scale, a set of 3122 lincRNAs defined by cDNA sequences was shown to be evolutionarily conserved relative to genomically neighbouring, presumed neutral, sequence [16]. lincRNA promoters are particularly highly constrained, suggesting that it could be the act of transcription itself that is more often important for these lincRNAs' functionality [16,23]. A similar pattern was observed in a second set of 1675 mouse lincRNAs which was defined using chromatin markers for active promoters (H3K4me3) and actively transcribed exonic sequence (H3K36me3) [12], and also in a collection of 1119 lincRNAs in *D. melanogaster* that were discovered using RNA-Seq transcriptome evidence [9].

Analysis of expression patterns

Instead of considering sequence conservation, other approaches have considered the conservation of transcription to be an indicator of these transcripts' functionality. A detailed analysis of four mouse lincRNAs revealed that their brain expression patterns can be conserved between diverse vertebrates, such as chicken and opossum, a marsupial [24]. lincRNAs can also be identified between more diverse species where their position, if not the primary sequence, is conserved. Comparisons between *Drosophila* and mouse have revealed an excess of these positionally equivalent lincRNAs [9]. This was also seen when comparing lincRNAs in zebrafish and humans, where the mutant phenotype of two of these zebrafish lincRNAs could even be rescued by the mature form of the positionally conserved mouse or human sequence [25]. Few lincRNAs are, however, so deeply conserved in their expression profiles. Of a sample of eight mouse lincRNAs validated by Northern blotting, five were also found to be expressed in rat, but none were found to be expressed in any of the human tissues or cell lines tested [26].

Another proposed indicator of functionality is when lincRNAs exhibit differential gene expression levels in different tissues or at different timepoints. It is argued that variations in gene expression levels reflect transcriptional regulation. For example, a survey of *in situ* hybridization images from the adult mouse brain collected by the Allen Mouse Brain Atlas revealed 849 lincRNAs that are expressed in the brain, 513 of which showed distinct regional patterns of expression [27]. Nevertheless, differences in transcript abundance might reflect inconsequential transcriptional events resulting from tissue- or developmental-stage-specific transcription factor binding and thus might not be considered to provide strong indicators of functionality.

lncRNA molecular mechanisms

Despite the difficulties inherent in their experimental investigation (see above), most of the few lncRNAs, including lincRNAs, that have been studied in detail have demonstrated roles in the regulation of gene expression. This regulation can be exerted at any one of a number of different levels, as illustrated in Figure 2.

Dosage compensation

Individual lincRNAs that aid in the regulation of chromosome-wide gene expression have been identified in both *Drosophila* and mammals. Specific lincRNAs are involved in dosage

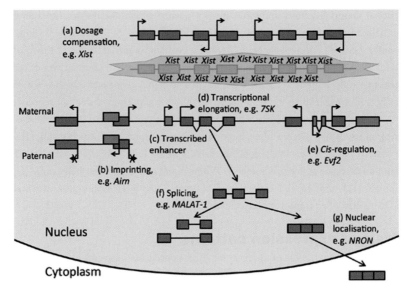

Figure 2. Schematic diagram describing lncRNA functions within a mammalian female cell

Protein-coding genes are shown in red, and lncRNA loci are in green. An arrow indicates the directionality of active transcription, whereas crossed-out arrows represent transcriptionally silenced alleles. *Airn* and *Evf2* loci are illustrated here and in the genome browser screenshots of Figure 1. Note that this diagram is not drawn to scale. (**a**) Silencing of gene expression along a complete chromosome, e.g. dosage compensation by *Xist* 'coating' the X chromosome. (**b**) Regulation of imprinted gene expression from one allele, e.g. *Airn*. (**c**) Transcription from upstream enhancer region. (**d**) Modulation of transcriptional elongation, e.g. *7SK*. (**e**) Regulation of genomically adjacent protein-coding genes, e.g. *Evf2*. (**f**) Interaction with splicing factors, e.g. *MALAT-1*. (**g**) Regulation of NFAT nuclear localization, e.g. *NRON*.

compensation which equalizes the dosage of gene expression from the X chromosome between females with two X chromosomes and males with only a single X chromosome. In *Drosophila*, this is achieved by hypertranscription from the single X chromosome in males, whereas, in mouse, transcription from one X chromosome is mostly inactivated in female cells.

In *Drosophila*, transcription from the male X chromosome is regulated by the MSL (male-specific lethal) complex, a complex containing several proteins [MSL1, MSL2, MSL3, MOF (Males absent on the first) and MLE (Maleless)] and two lincRNAs [RNA on X1 (*roX1*) and RNA on X2 (*roX2*)], which are both transcribed from the X chromosome [28]. Mutant analysis has suggested that the complex binds 30–40 'entry' sites on the X chromosome [29] and then spreads in *cis* to coat the entire chromosome, leading to H4K16 acetylation at actively transcribed gene loci, a more diffuse chromosome morphology and hypertranscription [30]. *roX1* and *roX2* are functionally redundant, despite sharing little sequence similarity and displaying distinct embryonic expression profiles [31]. These observations can be reconciled, at least in part, by experiments showing that most of the sequence of these lincRNAs is not required for normal function [31].

The equivalent mechanism in mice is quite different to that in *Drosophila*. Eutherian XCI (X chromosome inactivation) is thought to require the coating of the X chromosome by RNA produced from only one lincRNA locus, the 15 kb *Xist* (X-inactive specific transcript) [32]. This lincRNA is transcribed from the inactive X chromosome and spreads strictly in *cis* to coat the chromosome and prevent transcription [33]. Whether this requires a complex of proteins, as in *Drosophila*, or whether the act of coating by this lincRNA is sufficient to modify the X chromosome chromatin environment, remains unknown. The sequence of *Xist* is only 60% conserved across eutherian mammals, but the gene structure is conserved between mouse and humans, with several short well-conserved regions [34]. *Xist* expression is regulated by an antisense-encoded lincRNA, named *Tsix* [35], whose promoter is 13 kb downstream from *Xist*.

Imprinting

lncRNAs can also regulate the expression of genes that are physically linked on the chromosome, such as those present in imprinted gene clusters. Imprinted genes are expressed from only one allele in a diploid animal, and this expression depends on whether it is inherited from the maternal or paternal allele. Imprinted genes generally occur in clusters, suggesting that they are regulated as a single domain where lncRNA expression from one allele is often associated with repression of the protein-coding gene on that allele. One such example is the mouse lncRNA *Airn*, whose locus overlaps the *Igf2r* gene (Figure 1A). *Airn* is normally transcribed exclusively from the paternal allele, where it prevents expression of a gene cluster containing *Igf2r*, *Slc22a2* and *Slc22a3* from that allele, despite being antisense only to *Igf2r* [36]. Disrupting the *Airn* promoter causes *Igf2r* to be expressed from the paternal as well as the maternal allele [37]. The mechanism causing this remains unclear, although it appears to involve H3K9me3 recruitment at the *Slc22a3* promoter [38]. The *Airn* locus is poorly

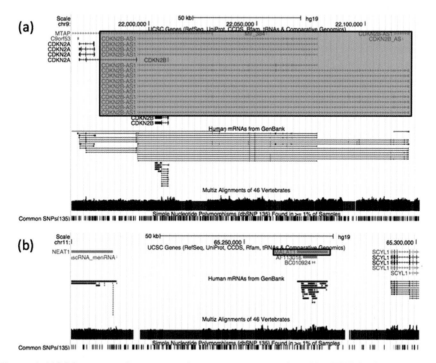

Figure 3. UCSC genome browsers of two cancer-associated lncRNA loci
The human lncRNAs and their supporting transcript models are highlighted in the yellow boxes.
(a) *ANRIL*. (b) *MALAT-1*.

conserved across related species [39], suggesting that this type of lncRNA regulation may be lineage-specific.

Cis- compared with *trans*-regulation

Recently, the question of how lincRNAs regulate individual protein-coding genes has attracted interest from several research groups. lincRNAs are thought to regulate gene expression either in *cis*, where the lincRNA acts only on the same DNA strand from which it is transcribed, or in *trans*, where the lincRNA can regulate genes located on other chromosomes.

For example, in mouse, the *Dlx5* and *Dlx6* protein-coding genes are up-regulated in cell culture by a lncRNA known as *Evf2* (Figure 1b) whose transcribed locus overlaps an intergenic enhancer lying between *Dlx5* and *Dlx6* [40]. The lncRNA and the two protein-coding transcripts share a similar expression pattern in the ventral forebrain, and *Evf2* can drive reporter expression in two neural cell lines in a dose-dependent manner. In mice expressing a truncated *Evf2* transcript, it appears that *Dlx5* and *Dlx6* are up-regulated in contrast with what was observed in cell culture [41], and these mice show a reduced number of GABAergic neurons in the early postnatal hippocampus. This effect on the GABAergic neuron counts is unlikely to be mediated through *Dlx5* and *Dlx6*, suggesting that *Evf2* may also possess a *trans*-acting function.

Attempts have been made to identify examples of potential *cis*-regulation at a genomic scale. Mouse lincRNAs tend to be found transcribed near to genes annotated with specific Gene Ontology terms and, in particular, to genes also involved in the regulation of transcription [12,42]. The expression of several of these lincRNA-protein-coding gene pairs has been subsequently investigated experimentally by *in situ* hybridization [42]. They were shown to have similar expression patterns in the developing mouse brain, which suggests that the lincRNA may positively regulate the expression of its adjacent protein-coding gene. *Cis*-encoded RNAs have also been identified which are transcribed through gene enhancers and which may function in a positive manner to regulate target gene expression. Recent experiments on a set of these eRNAs (enhancer RNAs) in human cells have suggested that the mature RNA molecule can be essential. siRNA (small interfering RNA)-mediated knockdown of seven (of 12 tested) putative eRNAs resulted in a significant disruption of transcription of a genomically proximal protein-coding gene [43]. At the human growth hormone locus, it appears to be the act of transcription through the eRNA which is important, because its enhancing effect remains even when the complete RNA sequence is replaced [44].

Another study has, however, suggested that this type of *cis*-relationship may confer little or no functional relevance, as siRNA constructs targeting a set of 147 lincRNAs in mouse embryonic stem cells revealed mostly *trans*-acting effects of disrupting the expression of these lincRNAs [45]. This type of experiment targets the mature RNA molecule and preferentially reveals the *trans*-acting functions of these lincRNAs. Whether lincRNAs act to regulate protein-coding gene transcription largely in *cis* or in *trans* remains a focus of much current research.

Transcription elongation

A lincRNA has been defined which, rather than regulating the initiation of transcription, negatively regulates elongation of transcription beyond the gene promoter. *7SK*, a 331 nt lincRNA, is thought to be involved in this process by sequestering proteins involved in transcriptional elongation, such as P-TEFb (positive transcription elongation factor) components, away from transcribed sites. Instead, they are stored in restricted domains, known as speckles, within the nucleus [46]. Recently, a possible *Drosophila* orthologue of *7SK* has also been identified, suggesting that this mechanism may be deeply conserved across evolution [47]. The downstream genes that are affected by this remain to be identified.

Post-transcriptional regulation

lincRNAs have also been discovered that are involved in gene regulation beyond the act of their transcription. In *D. melanogaster*, three non-coding heat–shock response (hsr-ω) transcripts are induced from the 93D locus by heat shock, CO_2 exposure and following the release of ecdysone in third instar larvae [48]. This 10–20 kb locus is functionally conserved in all Drosophilid species. One short transcript is cytoplasmic, whereas the other two remain at the locus from which they are transcribed and within nuclear 'omega' speckles thought to be storage sites for RNA-processing proteins. Although they are up-regulated in response to stress, these transcripts must also play a housekeeping role in the developing animal as only 20–25% of trans-heterozygote mutant embryos hatch [49].

Protein function can also be regulated by lincRNAs, as exemplified by the ncRNA repressor of the nuclear factor of activated T-cells (*NRON*). This lincRNA contains three exons and is alternatively spliced to produce transcripts varying in length from 800 nt to 3.7 kb [50]. Through its interaction with the nuclear import factor KPNB1, *NRON* contributes to preventing the protein product of *NFAT* (nuclear factor of activated T-cells) from entering the nucleus which, in turn, prevents NFAT from promoting transcription of its target genes [50].

lincRNAs have been implicated in diverse human diseases

lincRNAs are now known to be involved in a diverse array of cellular processes, examples of which were discussed above. This suggests that, when deleterious alleles occur within their loci, then an abnormal phenotype may arise. In humans, this might be manifested in disease. In fact, lincRNAs may be involved in various nervous system diseases through their association with protein-coding genes which are important in diseases such as Fragile X syndrome and Alzheimer's disease (reviewed in [51]). Several lincRNAs have also been implicated in cancer, which might be expected as their principal role appears to be in regulating gene expression, a dysregulation of which is a hallmark of tumorigenesis. In the following section, we discuss two specific loci as examples of disease-associated lncRNAs, *ANRIL* (CDKN2B antisense RNA 1) and *MALAT-1* (metastasis associated in lung adenocarcinoma transcript-1), which are proposed as being involved in the pathogenesis of different, but overlapping, types of cancer (Figure 3).

ANRIL

ANRIL is a 3834 nt RNA molecule made up of 19 exons that is transcribed from within the 9p21.3 locus. This locus is associated with susceptibility to a variety of complex diseases, including coronary artery disease, ischaemic stroke, aortic aneurysm, Type II diabetes, glioma, several carcinomas, malignant melanoma and acute lymphoblastic leukaemia [52]. The 9p21.3 locus also contains two cyclin-dependent kinase inhibitor genes, *CDKN2A* and *CDKN2B*.

Among several SNVs (single nucleotide variants) that have been associated with disease, many are more strongly associated with *ANRIL* than with *CDKN2A* or *CDKN2B* expression

[52]. *ANRIL* interacts with at least two protein complexes (Polycomb repressive complex 1 and 2) which negatively regulate the expression of *CDKN2A* and *CDKN2B* respectively [53,54]. Furthermore, the regulation of *CDKN2B* may act in *cis*, as suggested by the stronger response of an exogenous reporter, relative to the endogenous gene, when inserted near to an antisense construct [55]. It has been speculated that this regulation of other genes by ANRIL may contribute to cellular aging and, thereby, its involvement in several, seemingly unrelated diseases.

MALAT-1

MALAT-1 expression is found in a broad range of tissues and is significantly greater in metastatic NSCLC (non–small cell lung cancer) than in tumours which did not metastasize [56]. High expression of *MALAT-1* is also related to a reduced survival rate of patients with stage I NSCLC [56]. Since this initial study, *MALAT-1* has been found to be similarly up-regulated in other carcinomas, including breast, pancreas and colon cancer [57], which suggests that *MALAT-1* may have a general importance in carcinogenesis. *MALAT-1* transcripts are *trans*-acting as they are retained in the nucleus, where they influence alternative splicing of hundreds of transcripts by regulating the localization to nuclear speckles of multiple pre-mRNA splicing factors such as the serine/arginine splicing factor SF1 [58].

Conclusions

Despite only gaining widespread recognition relatively recently, it is becoming increasingly clear that large numbers of lncRNAs are transcribed from virtually all eukaryotic genomes sequenced to date. Detailed studies of individual loci have revealed that these genes can contribute to a wide range of cellular and molecular phenotypes. Many of the lncRNAs identified to date have been implicated in the regulation of gene expression, yet even the few examples that could be presented in this brief review show the diversity of cellular processes with which lncRNAs have been associated.

Increasing access to relatively low-cost next-generation sequencing will allow identification of large numbers of hitherto unseen lincRNA loci expressed in a variety of tissues (owing to their high tissue-specificity), in many different species and under a variety of physiological conditions. The future challenge of lincRNA biology thus lies less in the initial identification of lincRNAs, but in the determination of their relative contributions to different cellular processes and to lineage-specific biology.

Summary
- Thousands of lncRNA (long non-coding RNA) loci have already been identified in a variety of species, suggesting that they are important components of metazoan genomes.
- Expressed sequence tag and RNA-Seq analysis can be used to identify lncRNAs as novel transcripts with little or no protein-coding potential.
- Genome-wide evidence for lncRNA function has been provided by analyses of their evolutionary constraint and, to a lesser extent, differential expression profiles.

- Most lincRNAs (long intergenic non-coding RNAs) which have been studied in detail appear to be involved in the regulation of transcript abundance at some level, from chromatin modification to nuclear localization.
- Future work is likely to focus on larger-scale experimental analysis of lncRNA function.

Work of the authors is supported by the UK Medical Research Council and by an ERC Advanced Grant.

References

1. Johnson, J.M., Edwards, S., Shoemaker, D. and Schadt, E.E. (2005) Dark matter in the genome. Evidence of widespread transcription detected by microarray tiling experiments. Trends Genet. **21**, 93–102
2. Bertone, P., Stolc, V., Royce, T.E., Rozowsky, J.S., Urban, A.E., Zhu, X., Rinn, J.L., Tongprasit, W., Samanta, M., Weissman, S. et al. (2004) Global identification of human transcribed sequences with genome tiling arrays. Science **306**, 2242–2246
3. Stolc, V., Gauhar, Z., Mason, C., Halasz, G., Van Batenburg, M.F., Rifkin, S.A., Hua, S., Herreman, T., Tongprasit, W., Barbano, P.E. et al. (2004) A gene expression map for the euchromatic genome of *Drosophila melanogaster*. Science **306**, 655–660
4. van Bakel, H., Nislow, C., Blencowe, B.J. and Hughes, T.R. (2010) Most "dark matter" transcripts are associated with known genes. PLoS Biol. **8**, e1000371
5. Clark, M.B., Amaral, P.P., Schlesinger, F.J., Dinger, M.E., Taft, R.J., Rinn, J.L., Ponting, C.P., Stadler, P.F., Morris, K.V., Morillon, A. et al. (2011) The reality of pervasive transcription. PLoS Biol. **9**, e1000625
6. Kowalczyk, M.S., Higgs, D.R. and Gingeras, T.R. (2012) Molecular biology, RNA discrimination. Nature **482**, 310–311
7. Ponting, C.P., Oliver, P.L. and Reik, W. (2009) Evolution and functions of long noncoding RNAs. Cell **136**, 629–641
8. Manak, J.R., Dike, S., Sementchenko, V., Kapranov, P., Biemar, F., Long, J., Cheng, J., Bell, I., Ghosh, S., Piccolboni, A. and Gingeras, T.R. (2006) Biological function of unannotated transcription during the early development of *Drosophila melanogaster*. Nat. Genet. **38**, 1151–1158
9. Young, R.S., Marques, A.C., Tibbit, C., Haerty, W., Bassett, A.R., Liu, J.L. and Ponting, C.P. (2012) Identification and properties of 1,119 candidate lincRNA loci in the *Drosophila melanogaster* genome. Genome Biol. Evol. **4**, 427–442
10. Cabili, M.N., Trapnell, C., Goff, L., Koziol, M., Tazon-Vega, B., Regev, A. and Rinn, J.L. (2011) Integrative annotation of human large intergenic noncoding RNAs reveals global properties and specific subclasses. Genes Dev. **25**, 1915–1927
11. Derrien, T., Johnson, R., Bussotti, G., Tanzer, A., Djebali, S., Tilgner, H., Guernec, G., Martin, D., Merkel, A., Knowles, D.G. et al. (2012) The GENCODE v7 catalog of human long noncoding RNAs. Analysis of their gene structure, evolution, and expression. Genome Res. **22**, 1775–1789
12. Guttman, M., Amit, I., Garber, M., French, C., Lin, M.F., Feldser, D., Huarte, M., Zuk, O., Carey, B.W., Cassady, J.P. et al. (2009) Chromatin signature reveals over a thousand highly conserved large non–coding RNAs in mammals. Nature **458**, 223–227
13. Khalil, A.M., Guttman, M., Huarte, M., Garber, M., Raj, A., Rivea Morales, D., Thomas, K., Presser, A., Bernstein, B.E., van Oudenaarden, A. et al. (2009) Many human large intergenic noncoding RNAs associate with chromatin-modifying complexes and affect gene expression. Proc. Natl. Acad. Sci. U.S.A. **106**, 11667–11672

14. Kapranov, P., St Laurent, G., Raz, T., Ozsolak, F., Reynolds, C.P., Sorensen, P.H., Reaman, G., Milos, P., Arceci, R.J., Thompson, J.F. and Triche TJ (2010) The majority of total nuclear–encoded non-ribosomal RNA in a human cell is 'dark matter' un-annotated RNA. BMC Biol. **8**, 149
15. Taft, R.J., Kaplan, C.D., Simons, C. and Mattick, J.S. (2009) Evolution, biogenesis and function of promoter-associated RNAs. Cell Cycle **8**, 2332–2338
16. Ponjavic, J., Ponting, C.P. and Lunter, G. (2007) Functionality or transcriptional noise? Evidence for selection within long noncoding RNAs. Genome Res. **17**, 556–565
17. Okazaki, Y., Furuno, M., Kasukawa, T., Adachi, J., Bono, H., Kondo, S., Nikaido, I., Osato, N., Saito, R., Suzuki, H. et al. (2002) Analysis of the mouse transcriptome based on functional annotation of 60,770 full-length cDNAs. Nature **420**, 563–573
18. Kong, L., Zhang, Y., Ye, Z.Q., Liu, X.Q., Zhao, S.Q., Wei, L. and Gao, G. (2007) CPC, assess the protein-coding potential of transcripts using sequence features and support vector machine. Nucleic Acids Res. **35**, W345–W349
19. Lin, M.F., Jungreis, I. and Kellis, M. (2011) PhyloCSF, a comparative genomics method to distinguish protein coding and non-coding regions. Bioinformatics **27**, i275–i282
20. Wilson, B.A. and Masel, J. (2011) Putatively noncoding transcripts show extensive association with ribosomes. Genome Biol. Evol. **3**, 1245–1252
21. Dinger, M.E., Pang, K.C., Mercer, T.R. and Mattick, J.S. (2008) Differentiating protein-coding and noncoding RNA. Challenges and ambiguities. PLoS Comput. Biol. **4**, e1000176
22. Washietl, S., Pedersen, J.S., Korbel, J.O., Stocsits, C., Gruber, A.R., Hackermüller, J., Hertel, J., Lindemeyer, M., Reiche, K., Tanzer, A. et al. (2007) Structured RNAs in the ENCODE selected regions of the human genome. Genome Res. **17**, 852–864
23. Carninci, P., Kasukawa, T., Katayama, S., Gough, J., Frith, M.C., Maeda, N., Oyama, R., Ravasi, T., Lenhard, B., Wells, C. et al. (2005) The transcriptional landscape of the mammalian genome. Science **309**, 1559–1563
24. Chodroff, R.A., Goodstadt, L., Sirey, T.M., Oliver, P.L., Davies, K.E., Green, E.D., Molnár, Z. and Ponting, C.P. (2010) Long noncoding RNA genes, conservation of sequence and brain expression among diverse amniotes. Genome Biol. **11**, R72
25. Ulitsky, I., Shkumatava, A., Jan, C.H., Sive, H. and Bartel, D.P. (2011) Conserved function of lincRNAs in vertebrate embryonic development despite rapid sequence evolution. Cell **147**, 1537–1550
26. Babak, T., Blencowe, B.J. and Hughes, T.R. (2005) A systematic search for new mammalian noncoding RNAs indicates little conserved intergenic transcription. BMC Genomics **6**, 104
27. Mercer, T.R., Dinger, M.E., Sunkin, S.M., Mehler, M.F. and Mattick, J.S. (2008) Specific expression of long noncoding RNAs in the mouse brain. Proc. Natl. Acad. Sci. U.S.A. **105**, 716–721
28. Amrein, H. and Axel, R. (1997) Genes expressed in neurons of adult male *Drosophila*. Cell **88**, 459–469
29. Kelley, R.L., Meller, V.H., Gordadze, P.R., Roman, G., Davis, R.L. and Kuroda, M.I. (1999) Epigenetic spreading of the *Drosophila* dosage compensation complex from roX RNA genes into flanking chromatin. Cell **98**, 513–522
30. Deng, X. and Meller, V.H. (2006) Non-coding RNA in fly dosage compensation. Trends Biochem. Sci. **31**, 526–532
31. Meller, V.H. and Rattner, B.P. (2002) The roX genes encode redundant male-specific lethal transcripts required for targeting of the MSL complex. EMBO J. **21**, 1084–1091
32. Brockdorff, N., Ashworth, A., Kay, G.F., McCabe, V.M., Norris, D.P., Cooper, P.J., Swift, S. and Rastan, S. (1992) The product of the mouse Xist gene is a 15 kb inactive X-specific transcript containing no conserved ORF and located in the nucleus. Cell **71**, 515–526
33. Penny, G.D., Kay, G.F., Sheardown, S.A., Rastan, S. and Brockdorff, N. (1996) Requirement for Xist in X chromosome inactivation. Nature **379**, 131–137
34. Nesterova, T.B., Slobodyanyuk, S.Y., Elisaphenko, E.A., Shevchenko, A.I., Johnston, C., Pavlova, M.E., Rogozin, I.B., Kolesnikov, N.N., Brockdorff, N. and Zakian, S.M. (2001) Characterization of the genomic Xist locus in rodents reveals conservation of overall gene structure and tandem repeats but rapid evolution of unique sequence. Genome Res. **11**, 833–849

35. Lee, J.T., Davidow, L.S. and Warshawsky, D. (1999) Tsix, a gene antisense to Xist at the X-inactivation centre. Nat. Genet. **21**, 400–404
36. Sleutels, F., Zwart, R. and Barlow, D.P. (2002) The non-coding Air RNA is required for silencing autosomal imprinted genes. Nature **415**, 810–813
37. Wutz, A., Smrzka, O.W., Schweifer, N., Schellander, K., Wagner, E.F. and Barlow, D.P. (1997) Imprinted expression of the Igf2r gene depends on an intronic CpG island. Nature **389**, 745–749
38. Nagano, T., Mitchell, J.A., Sanz, L.A., Pauler, F.M., Ferguson-Smith, A.C., Feil, R. and Fraser, P. (2008) The Air noncoding RNA epigenetically silences transcription by targeting G9a to chromatin. Science **322**, 1717–1720
39. Pang, K.C., Frith, M.C. and Mattick, J.S. (2006) Rapid evolution of noncoding RNAs. Lack of conservation does not mean lack of function. Trends Genet. **22**, 1–5
40. Feng, J., Bi, C., Clark, B.S., Mady, R., Shah, P. and Kohtz, J.D. (2006) The Evf-2 noncoding RNA is transcribed from the Dlx-5/6 ultraconserved region and functions as a Dlx-2 transcriptional coactivator. Genes Dev **20**, 1470–1484
41. Bond, A.M., Vangompel, M.J.W., Sametsky, E.A., Clark, M.F., Savage, J.C., Disterhoft, J.F. and Kohtz, J.D. (2009) Balanced gene regulation by an embryonic brain ncRNA is critical for adult hippocampal GABA circuitry. Nat. Neurosci. **12**, 1020–1027
42. Ponjavic, J., Oliver, P.L., Lunter, G. and Ponting, C.P. (2009) Genomic and transcriptional co-localization of protein-coding and long non-coding RNA pairs in the developing brain. PLoS Genetics **5**, e1000617
43. Ørom, U.A., Derrien, T., Beringer, M., Gumireddy, K., Gardini, A., Bussotti, G., Lai, F., Zytnicki, M., Notredame, C., Huang, Q. et al. (2010) Long noncoding RNAs with enhancer-like function in human cells. Cell **143**, 46–58
44. Yoo, E.J., Cooke, N.E. and Liebhaber, S.A. (2012) An RNA-independent linkage of non-coding transcription to long-range enhancer function. Mol. Cell. Biol. **32**, 2020–2029
45. Guttman, M., Donaghey, J., Carey, B.W., Garber, M., Grenier, J.K., Munson, G., Young, G., Lucas, A.B., Ach, R., Bruhn, L. et al. (2011) lincRNAs act in the circuitry controlling pluripotency and differentiation. Nature **477**, 295–300
46. Prasanth, K.V., Camiolo, M., Chan, G., Tripathi, V., Denis, L., Nakamura, T., Hübner, M.R. and Spector, D.L. (2010) Nuclear organization and dynamics of 7SK RNA in regulating gene expression. Mol. Biol. Cell **21**, 4184–4196
47. Nguyen, D., Krueger, B.J., Sedore, S.C., Brogie, J.E., Rogers, J.T., Rajendra, T.K., Saunders, A., Matera, A.G., Lis, J.T., Uguen, P. and Price, D.H. (2012) The *Drosophila* 7SK snRNP and the essential role of dHEXIM in development. Nucleic Acids Res. **40**, 5283–5297
48. Lakhotia, S.C. and Sharma, A. (1996) The 93D (hsr-omega) locus of *Drosophila*, non-coding gene with house-keeping functions. Genetica **97**, 339–348
49. Mohler, J. and Pardue, M.L. (1982) Deficiency mapping of the 93D heat-shock locus in *Drosophila melanogaster*. Chromosoma **86**, 457–467
50. Willingham, A.T., Orth, A.P., Batalov, S., Peters, E.C., Wen, B.G., Aza-Blanc, P., Hogenesch, J.B. and Schultz, P.G. (2005) A strategy for probing the function of noncoding RNAs finds a repressor of NFAT. Science **309**, 1570–1573
51. Qureshi, I.A., Mattick, J.S. and Mehler, M.F. (2010) Long non-coding RNAs in nervous system function and disease. Brain Res. **1338**, 20–35
52. Cunnington, M.S., Santibanez Koref, M., Mayosi, B.M., Burn, J. and Keavney, B. (2010) Chromosome 9p21 SNPs associated with multiple disease phenotypes correlate with ANRIL expression. PLoS Genet. **6**, e1000899
53. Yap, K.L., Li, S., Muñoz-Cabello, A.M., Raguz, S., Zeng, L., Mujtaba, S., Gil, J., Walsh, M.J. and Zhou, M.M. (2010) Molecular interplay of the noncoding RNA ANRIL and methylated histone H3 lysine 27 by polycomb CBX7 in transcriptional silencing of INK4a. Mol. Cell **38**, 662–674
54. Kotake, Y., Nakagawa, T., Kitagawa, K., Suzuki, S., Liu, N., Kitagawa, M. and Xiong, Y. (2011) Long non-coding RNA ANRIL is required for the PRC2 recruitment to and silencing of p15(INK4B) tumor suppressor gene. Oncogene **30**, 1956–1962

55. Yu, W., Gius, D., Onyango, P., Muldoon–Jacobs, K., Karp, J., Feinberg, A.P. and Cui, H. (2008) Epigenetic silencing of tumour suppressor gene p15 by its antisense RNA. Nature **451**, 202–206
56. Ji, P., Diederichs, S., Wang, W., Böing, S., Metzger, R., Schneider, P.M., Tidow, N., Brandt, B., Buerger, H., Bulk, E. et al. (2003) MALAT-1, a novel noncoding RNA, and thymosin β4 predict metastasis and survival in early-stage non-small cell lung cancer. Oncogene **22**, 8031–8041
57. Lin, R., Maeda, S., Liu, C., Karin, M. and Edgington, T.S. (2007) A large noncoding RNA is a marker for murine hepatocellular carcinomas and a spectrum of human carcinomas. Oncogene **26**, 851–858
58. Tripathi, V., Ellis, J.D., Shen, Z., Song, D.Y., Pan, Q., Watt, A.T., Freier, S.M., Bennett, C.F., Sharma, A., Bubulya, P.A. et al. (2010) The nuclear–retained noncoding RNA MALAT1 regulates alternative splicing by modulating SR splicing factor phosphorylation. Mol. Cell **39**, 925–938

10

Therapeutic targeting of non-coding RNAs

Thomas C. Roberts and Matthew J.A. Wood[1]

Department of Physiology, Anatomy and Genetics, University of Oxford, South Parks Road, Oxford OX1 3QX, U.K.

Abstract

ncRNAs (non-coding RNAs) are implicated in a wide variety of cellular processes, including the regulation of gene expression. In the present chapter we consider two classes of ncRNA: miRNAs (microRNAs) which are post-transcriptional regulators of gene expression and lncRNAs (long ncRNAs) which mediate interactions between epigenetic remodelling complexes and chromatin. Mutation and misexpression of ncRNAs have been implicated in many disease conditions and, as such, pharmacological modulation of ncRNAs is a promising therapeutic approach. miRNA activity can be antagonized with antisense oligonucleotides which sequester or degrade mature miRNAs, and expressed miRNA sponges which compete with target transcripts for miRNA binding. Conversely, synthetic or expressed miRNA mimics can be used to treat a deficiency in miRNA expression. Similarly, conventional antisense technologies can be used to silence lncRNAs. Targeting promoter-associated RNAs with siRNAs (small interfering RNAs) results in recruitment of chromatin-modifying activities and induces transcriptional gene silencing. Alternatively, targeting natural antisense transcripts with siRNAs or antisense oligonucleotides can abrogate endogenous epigenetic silencing leading to transcriptional gene activation. The ability to modulate gene expression at the epigenetic level presents exciting new opportunities for the treatment of human disease.

Keywords:

long non-coding RNA, microRNA, natural antisense transcript, transcriptional activation, transcriptional gene silencing.

[1]To whom correspondence should be addressed (email matthew.wood@dpag.ox.ac.uk).

Introduction

Transcription of the mammalian genome is ubiquitous [1,2] and occurs in both sense and antisense orientations [3–5]. Only a small fraction (~1%) of the genome codes for protein and so the vast majority of cellular transcriptional output is therefore ncRNA (non-coding RNA). Although there are many classes of ncRNAs with wide-ranging functionalities (e.g. RNA editing, mediation of mRNA splicing, ribosomal function), in the present chapter we will primarily be concerned with the role of ncRNAs in the regulation of gene expression. As modulators of gene expression, ncRNAs are promising therapeutic targets. In the present chapter we will consider two general classes of ncRNA molecules: miRNAs (microRNAs) and lncRNAs (long ncRNAs).

miRNAs

miRNAs are small RNA sequences (21–23 nt) that are endogenous RNAi (RNA interference) effectors [6]. Typically, miRNAs are derived from longer pri-miRNA (primary-miRNA) transcripts that are transcribed by RNA Pol II (RNA polymerase II). Pri-miRNAs can either be intergenic transcripts or the mRNAs of protein-coding genes (with the miRNA sequences contained within one or more introns). Pri-miRNA transcripts are progressively processed by the enzyme Drosha (in the nucleus) and then Dicer (in the cytoplasm) to generate the pre-miRNA (precursor-miRNA) hairpin and mature miRNA species respectively. Following Dicer cleavage, one strand of the miRNA hairpin is loaded into the RNA-binding protein AGO2 (Argonaute 2), the so called 'catalytic engine of RNAi', and a component of the RISC (RNA-induced silencing complex) [7]. The mature miRNA sequence guides the RISC to its mRNA targets in the cytoplasm where it binds, typically in the 3' UTR (untranslated) region, to form an imperfect duplex and induce translational repression and/or mRNA decay [8–10] (Figure 1). As miRNAs can induce silencing with only partial complementarity to a target transcript (i.e. multiple mRNA–miRNA mismatches are tolerated), each miRNA can bind to multiple mRNA targets. Similarly, each mRNA 3' UTR contains multiple potential miRNA-binding sites. Consequently miRNAs can act as master regulators of gene expression by regulating families of transcripts with related functions [11]. Individual miRNAs have been implicated in a wide variety of physiological and pathophysiological processes and, as such, are potential pharmacological targets [12].

miRNA therapeutics

In instances where the activity of an miRNA is a causitive factor in pathology, strategies which antagonize miRNA activity are desirable. These fall into two broad categories: (i) small oligonucleotide miRNA inhibitors and (ii) expressed miRNA sponges. Conversely, miRNA replacement therapy can be utilized to correct an miRNA deficiency, or to modulate an endogenous protective pathway (Figure 1).

miRNA antagonism

There are numerous examples of miRNAs that promote pathology in human disease and are consequently therapeutic targets. For example, many viruses express specific miRNA genes

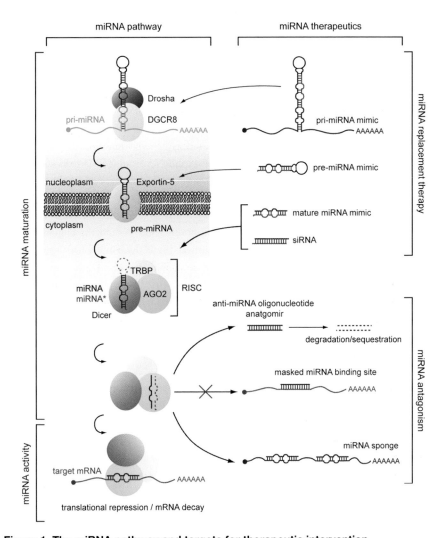

Figure 1. The miRNA pathway and targets for therapeutic intervention
Schematic diagram showing the miRNA pathway whereby pri-miRNA transcripts are progressively processed to generate the mature miRNA species. First, a complex of Drosha and DGCR8 (Di George Syndrome critical region gene 8) cleaves the pri-miRNA transcript to produce the pre-miRNA in the nucleus. Export of the pre-miRNA hairpin from the nucleus is facilitated by exportin-5. In the cytoplasm, the pre-miRNA is cleaved by Dicer generating a ~21 bp RNA duplex. Dicer is part of the RISC along with AGO2 and TAR RNA-binding protein 2 (TRBP). One strand of the duplex is loaded into AGO2 and the other strand (labelled miRNA*) is subsequently degraded. RISC is then guided to complementary target mRNAs and induces translational repression or mRNA decay. Approaches for miRNA replacement therapy and miRNA antagonism are indicated. miRNA mimics enter the miRNA pathway at various stages. Expressed pri-miRNA and pre-miRNA mimics enter at the Drosha and Dicer cleavage steps respectively. Conversely, mature miRNA mimics (either bulged duplexes or completely complementary siRNAs) enter RISC directly without prior processing. Anti-miRNA oligonucleotides or antagomirs bind to mature miRNA species in the cytoplasm and induce degradation or sequestration of the target miRNA. Similarly, oligonucleotides can bind to target mRNAs and mask an miRNA-binding site. Expressed miRNA sponges are transcripts containing multiple miRNA-binding sites which compete with endogenous mRNAs for miRNA binding.

[13], or are dependent upon host cell miRNAs for viral replication [14]. One promising application of miRNA inhibition is in the treatment of cancer, as miRNA-mediated gene regulation has been implicated in tumorigenesis and metastasis [15]. For example, inhibition of miR-10b in a mouse mammary tumour model resulted in a reduction in lung metastasis [16]. Anti-miRNA technologies are currently the most advanced miRNA-based therapeutic strategies with the most commonly used approach being AMOs (anti-miRNA oligonucleotides). These are single-stranded oligonucleotides consisting of the reverse complement sequence of a target miRNA that function by either degrading or sequestering the target miRNA. Alternatively, oligonucleotides complementary to target mRNAs block miRNA binding at individual recognition sites [17]. This target masking strategy allows for the inhibition of specific miRNA–mRNA interactions.

The *in vivo* delivery of AMOs is a substantial obstacle to their effective use as therapeutics. Typically, AMOs contain extensive chemical modification to both the oligonucleotide backbone and the ribose sugar (e.g. substitution at the 2′-hydroxy with O-methyl or O-methoxyethyl groups) in order to increase nuclease stability, reduce clearance and improve pharmacokinetic properties. For example, the commonly used phosphorothioate backbone modification has the dual effect of improving oligonucleotide nuclease stability and conferring non-specific protein binding with a concomitant increase in serum half-life. The incorporation of LNA (locked nucleic acid) bases results in an increase in the T_m (melting temperature) of AMOs and favours binding to RNA over DNA [18]. Similarly, the nucleic acid analogue PNA (peptide nucleic acid) has also been utilized to antagonize miRNA activity [19,20]. Alternatively, the conjugation of AMOs with lipophilic moieties, such as cholesterol in the case of antagomirs, results in improved cellular uptake [21,22] (Figure 2). AMOs have successfully been delivered systemically to liver, kidney, bone marrow and adipose tissue, and locally delivered to lung, gut, brain and eye in model organisms (reviewed in [23]). It has also recently been shown that naturally derived lipid microvesicles called exosomes can deliver exogenous RNA cargos to brain following systemic administration [24]. It will be interesting to see whether AMOs can be delivered via a similar strategy.

At the time of writing the most advanced anti-miRNA therapy is currently in Phase IIa clinical trials (http://www.clinicaltrials.gov). Miravirsen, developed by Santaris Pharma A/S to treat chronic HCV (hepatitis C virus) infection, is a 15-mer LNA-modified phosphorothioate antisense oligonucleotide inhibitor of miR-122. Endogenous miR-122 is required for HCV viral replication [14] and antagonism of this miRNA was shown to reduce viraemia in a chronically infected chimpanzee model with no evidence of toxicity [25].

The effects of AMOs are transient as they are dependent on the presence of the effector molecule. Consequently, expressed miRNA decoys or sponges have been developed in order to elicit longer-term miRNA inhibition [26]. These virus or plasmid-encoded transcripts contain multiple miRNA target sites and compete with endogenous target mRNAs for miRNA binding. miRNA sponges have been used successfully to inhibit miR-9 in highly malignant 4T1 cells, leading to suppression of metastasis [27]. Expressed miRNA inhibitors are an alternative therapeutic modality to the use of AMOs although, as they will probably require viral vector-mediated delivery *in vivo*, they are subject to the limitations and risks associated with classical gene therapy [28].

miRNA inhibition strategies are not limited to diseases in which aberrant miRNA expression is a causative factor in the pathology. In the muscle wasting condition DMD

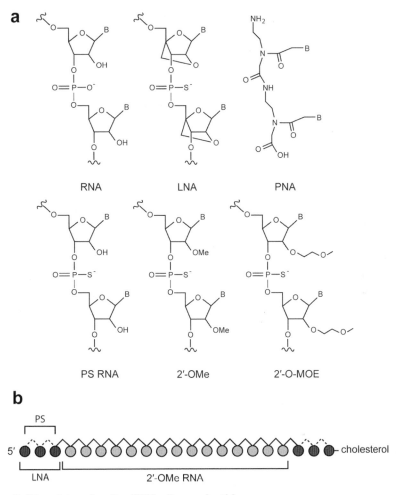

Figure 2. Chemistry of anti-miRNA oligonucleotides
(a) Examples of nucleic acid chemistries utilized in anti-miRNA oligonucleotides. RNA, LNA, PNA (peptide nucleic acid), PS RNA (phosphorothioate RNA), and RNA with 2′-OMe (O-methyl) or 2′-O-MOE (O-methoxyethyl) substitutions at the 2′ carbon of the ribose sugar. (b) Design of a generic antagomir. Nucleotides have O-methyl modifications at the 2′ position (light grey circles). The terminal nucleotides are LNA chemistry (dark grey circles) with phosphorothioate (PS) backbones (dotted lines). Additionally, the oligonucleotide has a cholesterol group conjugated to the 3′ terminus. LNA, 2′-OMe and 2′-O-MOE chemistries are commonly combined with phosphorothioate backbone modifications and so are depicted as such.

(Duchenne muscular dystrophy), loss-of-function mutations in the gene which encodes the dystrophin protein lead to progressive muscle weakness and are ultimately fatal. miR-31 is highly up-regulated in dystrophic muscle and acts to suppress translation of dystrophin mRNA. This interaction is clinically relevant as the activity of miR-31 limits the efficacy of efforts to restore dystrophin protein expression by exon skipping therapy. Inhibition of miR-31 using an miRNA sponge in combination with exon skipping was shown to be more effective at restoring dystrophin than exon skipping alone [29]. This study demonstrates that modulating miRNA activity can be an effective means of boosting the expression of therapeutically relevant genes.

miRNA replacement therapy

miRNA mimics are synthetic or expressed oligonucleotides that mimic the function of endogenous miRNAs. As miRNA expression is frequently dysregulated in tumours [30] and some miRNAs have been shown to have tumour-suppressive functionality [31], miRNA mimics are potential anti-cancer therapeutics. For example, adenovirus-mediated delivery of an expressed miR–26a mimic resulted in inhibition of tumour progression in a murine hepatocellular carcinoma model [32]. Conversely, miRNA mimics can be used to modulate pathophysiological processes. miR-29 is known to regulate fibrosis by suppressing the expression of collagens, fibrillins and elastin [33]. miR-29 expression is reduced in dystrophic muscle, leading to fibrogenesis in DMD. Consequently, treatment with a synthetic miR-29 mimic was capable of reducing fibrosis and improving pathology in the *mdx* mouse model of DMD [34].

Considering that AMOs are required only to bind mature single-stranded miRNAs with high affinity and specificity there are relatively few constraints on what chemical modifications can be incorporated in oligonucleotide design. Conversely, synthetic miRNA mimics have much more stringent requirements on chemical composition as RISC incorporation is essential for function. Synthetic miRNA mimics must primarily consist of RNA nucleotides with relatively few chemical modifications tolerated. As such, the development of therapeutic miRNA mimics has lagged behind that of anti-miRNA technology. Additionally, the delivery of synthetic miRNA mimics is subject to the same delivery obstacles as siRNAs (small interfering RNAs) (reviewed in [35]).

miRNA-mediated transgene inactivation

Classical gene therapy typically involves the delivery of DNA encoding a therapeutic transgene to treat or manage disease. Transgene expression in professional APCs (antigen-presenting cells) can lead to immune-related vector clearance and consequently limit therapeutic efficacy. To address this problem, Brown et al. [36] engineered a lentiviral vector expressing a GFP (green fluorescent protein) reporter to contain miRNA target sites complementary to miR-142-3p (which is highly expressed in immune cells) in its 3′ UTR. This strategy restricted GFP expression to non-haemopoietic cells and resulted in stable transgene expression in the desired target tissues [36]. A similar approach has been used with respect to adenovirus-mediated oncolytic virotherapy for cancer. In this case, high expression levels of the adenoviral E1A protein in hepatocytes results in acute liver toxicity. To abrogate this toxicity, miRNA target sites for a liver-specific miRNA, miR-122, were inserted into the 3′ UTR of the viral transgene. Mice treated with these miRNA-restricted adenoviral vectors showed reduced viral replication in the liver and almost no liver toxicity [37]. Thus, by taking advantage of endogenous miRNA regulation, the expression of therapeutic transgenes can be fine-tuned to minimize toxic off-target effects.

lncRNAs

lncRNAs are a heterogeneous group of RNA transcripts >200 nt in length with low protein-coding potential that are processed in similar ways to mRNAs (i.e. the majority are spliced and many are polyadenylated). lncRNAs can be sense or antisense transcripts with respect to a neighbouring protein-coding gene locus, intron-derived, the products of divergent bidirectional transcription or reside in the space between genes [i.e. lincRNAs (long intergenic ncRNAs)] (Figure 3a). Many lncRNAs show distinct spatial, temporal, cell-type specific and subcellular-specific expression patterns. It has been argued that this is indicative of tightly regulated transcription and therefore unlikely to be transcriptional 'noise' [38,39]. In addition, many lncRNAs show a high degree of evolutionary conservation consistent with biological function [40]. Given that there are thousands of lncRNAs, many of which have been implicated in cellular processes, these transcripts represent a multitude of potential drug targets.

lncRNA function

lncRNAs primarily act as adaptors that mediate interactions between chromatin, proteins and other RNAs. Two properties of lncRNA molecules enable them to function in this manner. First, they can bind to DNA or other RNA molecules by (i) complementary Watson–Crick base pairing to form hetero- or homo-duplexes, (ii) formation of DNA–DNA–RNA triplexes by Hoogstein and reverse Hoogstein base pairing, or (iii) by direct RNA recognition of chromatin surface features [41]. Secondly, the inherent flexibility of RNA permits the formation of complex secondary structures that can function as binding domains for proteins or small molecules (Figure 3b). The combination of these properties enables a much wider range of functions than is possible with miRNAs. Additionally, lncRNAs may also contain multiple binding modules allowing for complex multi-functional interactions.

Arguably the most exciting role of lncRNAs is as epigenetic regulators of gene expression. lncRNAs form riboprotein complexes where the ncRNA acts as a guide that targets

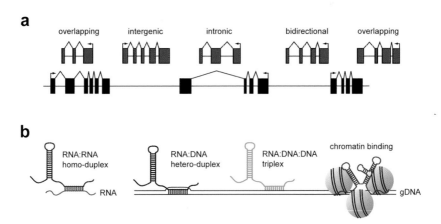

Figure 3. Genomic organization and function of lncRNAs
(**a**) Genomic organization of lncRNAs (blue) with respect to protein coding genes (black). Arrows indicate the direction of transcription. lncRNAs can be partially or completely overlapping with protein coding genes, reside within introns (intronic) or the space between genes (intergenic), or result from bidirectional transcription at promoters. (**b**) lncRNAs can interact with other RNA molecules (RNA–RNA homoduplex), transient single-stranded genomic DNA (gDNA) regions (RNA–DNA heteroduplex), form RNA–DNA–DNA triplexes with gDNA, or fold into secondary structures which can directly bind to chromatin.

chromatin-modifying activities or transcription factors to specific genomic loci. For example, RepA ncRNA [derived from the 5′ region of the *Xist* (X-inactive specific transcript) transcript] binds PRC2 (Polycomb Repressive Complex 2) and guides it to induce H3K27 (histone H3 Lys^{27}) tri-methylation and heterochromatin formation at the *Xist* promoter. This event initiates the process of XCI (X-chromosome inactivation) [42]. Similarly, in the case of rRNA genes, a promoter-associated RNA transcript forms a triplex structure at the rRNA promoter and recruits DNMT3B [DNA (cytosine-5-)-methyltransferase 3 β] in order to induce promoter DNA methylation and transcriptional gene silencing [43,44]. Thus RepA and the rDNA promoter transcript act as guides for epigenetic modifiers *in cis* (i.e. influencing neighbouring genes). Conversely, lncRNAs can also function as guides *in trans* (i.e. affecting distal genes). For example, lincRNA-p21 is able to guide epigenetic remodelling at multiple genomic loci via recruitment of hnRNP (heterogeneous nuclear ribonucleoprotein)-k [45].

lncRNAs can also act as RNA scaffolds that remain associated with a chromatin locus and recruit multiple epigenetic modifiers. As a result, complex changes in chromatin states can be dynamically co-ordinated in response to external signals. The lncRNA *HOTAIR*, which is transcribed from the *HOXC* cluster, binds to the *HOXD* cluster. HOTAIR associates with PRC2 at its 5′ region [46] and a complex containing LSD1, CoREST and REST at its 3′ region [47]. Consequently, the activities of these two complexes [i.e. tri-methylation of H3K27 and demethylation of H3K4 (histone H3 Lys^4) respectively] are co-ordinated in order to induce transcriptional silencing. Similarly, scaffold lncRNAs transcribed from pericentromeric satellite regions have been shown to associate with SUMOylated-HP1 (heterochromatin protein 1) which induces further recruitment of additional HP1 molecules and transcriptional silencing [48]. The epigenetic silencing complexes SMCX and PRC1, and the epigenetic activating complex WDR5/MLL have also been shown to associate with lncRNAs [49,50].

Recently, it was shown that lncRNAs can influence the subnuclear localization of genomic loci. The lncRNA TUG1 was shown to bind methylated Pc2 (a polycomb component) and thereby direct the accompanying genomic DNA to polycomb bodies where it is epigenetically silenced. Conversely, unmethylated Pc2 was bound by another lncRNA, MALAT1 (metastasis associated in lung adenocarcinoma transcript-1; also known as NEAT2), which resulted in localization to ICGs (interchromatin granules) which are associated with active transcription [51]. Thus post-translational modification of a non-histone protein is capable of influencing the subnuclear localization of chromatin through interactions with lncRNAs.

lncRNAs also act as 'riboregulator' decoys by binding and sequestering proteins (e.g. transcription factors and chromatin modifiers) or other RNAs. For example, the lncRNA Gas-5 (growth arrest-specific 5) forms an RNA secondary structure that binds to the DNA binding domain of GR (glucocorticoid receptor) and prevents it from interacting with its DNA target sites, thus repressing GR activity [52]. Additionally, lncRNAs have been shown to act as endogenous miRNA sponges [53,54] or to mask miRNA-binding sites on target mRNAs [55]. Similarly, MALAT1 alters the subnuclear localization of the splicing factor SR through sequestration in nuclear speckles [56,57].

lncRNAs also exert non-epigenetic effects on gene expression. Transcription of upstream lncRNA genes is sufficient, in some cases, to inhibit transcription of downstream genes, a phenomenon known as transcriptional interference [58]. In the case of the *DHFR* (dihydrofolate reductase) locus, a 5′ sense promoter RNA forms a direct association with *DHFR* promoter

DNA in order to inhibit transcription (in addition to acting as a decoy for the basal transcription factor TFIIB) [59].

Where overlapping bidirectional transcription occurs, there is potential for hybridization between the pair of sense and antisense transcripts. In the case of *BACE1* (β-secretase 1), a gene involved in the pathophysiology of Alzheimer's disease, a 3′-overlapping antisense RNA (BACE1-AS) forms an RNA duplex with the *BACE1* mRNA leading to stabilization and increased β-secretase expression [60,61]. Additionally, the formation of dsRNA (double-stranded RNA) as a result of overlapping transcription can form substrates for Dicer leading to the production of endo-siRNAs [endogenous-siRNAs (small interfering RNAs)] capable of silencing complementary target mRNAs [62,63]. Mechanisms of lncRNA action are summarized in Figure 4.

lncRNA in disease

Considering that lncRNAs are involved in a wide variety of gene regulation processes including epigenetic remodelling and memory, control of transcription and translation [60], cellular differentiation [64,65], XCI [66], mono-allelic expression of imprinted loci [67] and modulation of splicing [57] to name only a few, it is unsurprising that some lncRNAs are also implicated in human disease. Given that only a small portion of the genome (∼1%) codes for protein, the majority of mutations occur in non-coding regions [68]. Consequently, lncRNA mutations and misexpression have been linked to disease [69]. For example, mutations have been identified in lncRNAs transcribed from ultra-conserved regions in patients with CLL (chronic lymphocytic leukaemia) and CRC (colorectal cancer) [70]. Similarly, expression of HOTAIR is increased in breast tumours and correlates well with metastasis [71]. Increased levels of HOTAIR result in retargeting of the PRC2 to novel genomic sites with consequent changes in gene expression and increased tumour invasiveness. Similarly, MALAT1 was initially identified due to its association with lung metastasis [72].

In a landmark study by Tufarelli et al. [73] an antisense RNA was shown to be the direct cause of α-thalassaemia in a patient with a rare chromosome rearrangement. In this instance, the *LUC7L* promoter was found to be translocated immediately downstream of the *HBA2* (α-globin) gene resulting in transcription of a novel antisense RNA through the CpG island in the *HBA2* promoter. This antisense RNA mediates hypermethylation of the CpG island leading to epigenetic silencing of *HBA2* [73]. This study was the first to demonstrate how misexpression of a ncRNA could directly lead to human disease.

A study by Lewejohann et al. [74] revealed that mice in which the neuron-enriched ncRNA, *BC1*, was knocked-out showed no obvious physical or neurological defects relative to wild-type controls when under normal laboratory conditions. However, mutant mice exhibited reduced exploratory behaviour, increased anxiety and decreased survival upon reintroduction into a semi-natural outdoor environment [74]. This study raises the possibility that ncRNAs may also be involved in complex behavioural phenotypes in humans, specifically in disorders with cryptic aetiologies. Furthermore, the human homologue of *BC1*, *BC200*, is found to be up-regulated in the brains of Alzheimer's disease patients and is mislocalized to neuronal cell bodies instead of dendritic spines, suggesting that it may be involved in disease pathophysiology [75].

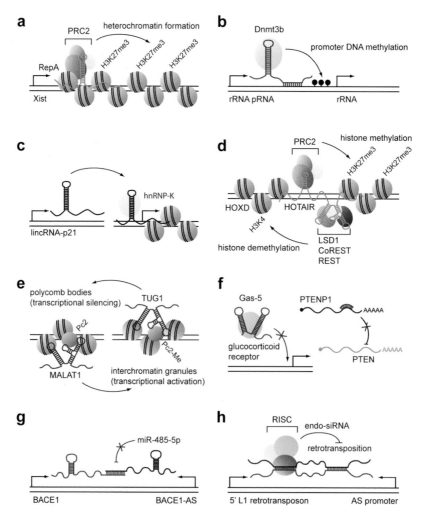

Figure 4. Mechanisms of gene regulation by lncRNAs
lncRNAs can act as guides for chromatin-modifying activities or transcription factors *in cis*. (**a**) RepA guides PRC2 to the Xist locus resulting in H3K27 tri-methylation and heterochromatin formation. (**b**) An rRNA promoter-associated RNA guides Dnmt3b to the rRNA promoter resulting in hypermethylation of the rRNA promoter and gene silencing. In addition rRNA pRNA also forms a triplex structure with the rRNA promoter. lncRNAs can also act as guides *in trans*. (**c**) lincRNA-p21 guides hnRNP-K to multiple sites in the genome and initiates gene silencing events. (**d**) lncRNAs can act as scaffolds for epigenetic-modifying complexes. The lncRNA HOTAIR binds both PRC2 and a complex containing LSD1, CoREST and REST to facilitate co-ordinated H3K27 tri-methylation and H3K4 demethylation at the HOXD locus. (**e**) Differential binding of lncRNAs TUG1 and MALAT1 to methylated or unmethylated Pc2 targets chromatin to polycomb bodies or interchromatin granules respectively. (**f**) lncRNAs can also act as riboregulator decoys for transcription factors (e.g. Gas-5 sequesters the glucocorticoid receptor) or miRNAs (e.g. the pseudogene PTENP1 acts as an miRNA sponge to relieve miRNA repression of PTEN). (**g**) Overlapping lncRNAs can hybridize with complementary mRNAs and stabilize them. In the case of the sense–antisense pair BACE1–BACE1-AS this is, at least in part, achieved by masking a binding site for miR-485-5p in the BACE1 3′ UTR. (**h**) Bidirectional transcription of lncRNAs can generate substrates for Dicer which are processed into endo-siRNAs. These can induce post-transcriptional gene silencing of complementary target RNAs (e.g. endo-siRNAs generated from bidirectional transcription of L1 retrotransposon sites leads to inhibition of retrotransposition).

© The Authors Journal compilation © 2013 Biochemical Society

Cross-talk between short and long RNAs

Whereas the targeting of mRNA sequences with antisense technologies (i.e. siRNAs and antisense oligonucleotides) for post-transcriptional gene silencing is now commonplace, in recent years a number of reports have demonstrated that targeting non-coding regions can also influence gene expression. Specifically, it is now apparent that transcription occurs at many promoters, enhancers and 3′ gene termini [76–78], and that these transcripts are targets for therapeutic modulation. Early studies targeting promoters with siRNAs reported both TGS (transcriptional gene silencing) [79–81] and TGA (transcriptional gene activation) [also known as RNAa (RNA activation)] [82–84]. Tentative mechanistic models have been proposed to explain these opposing phenomena [85].

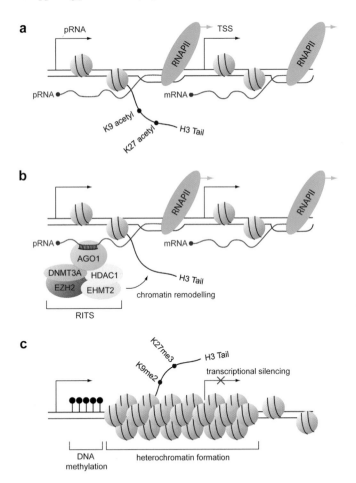

Figure 5. Model of small-RNA-mediated transcriptional gene silencing
(**a**) pRNA is transcribed from the promoter region of a hypothetical gene locus. This transcript remains tethered to the locus through association with RNA Pol II. (**b**) Targeting the pRNA with an siRNA recruits the RNA-induced transcriptional silencing complex (RITS) which includes AGO1, a histone deacetylase (HDAC1), histone methyltransferases (EZH2 and EHMT2) and a DNA methyl transferase (DNMT3A) leading to local chromatin remodelling. (**c**) Heterochromatin formation and promoter DNA methylation (black lollipops) results in transcriptional gene silencing of the targeted promoter.

In the case of TGS, the effector siRNA, shRNA or expressed antisense RNA is bound by AGO1 and directs the RITS (RNA-induced transcriptional silencing complex) to low-copy pRNAs (promoter-associated RNAs). RITS recruitment at a target promoter triggers remodelling of the local chromatin such that there is enrichment of silent-state chromatin modifications [i.e. H3K9me2 (histone H3 Lys^9 dimethylation) and H3K27me3 histone H3 Lys^{27} trimethylation)] and, in some cases, promoter DNA methylation. There is evidence that RITS is comprised of HDAC1 (histone deacetylase 1), the histone methyltransferases EZH2 and EHMT2 (formerly G9a), and DNMT3A [86,87] (Figure 5). pRNAs are sense orientation transcripts that are initiated upstream of the conventional transcription start site and are required for TGS [80]. They may act as *cis*-regulatory sequences that remain bound to the promoter chromatin or, alternatively, they may simply be tethered to the locus by association with RNA Pol II. There is evidence that unstable pRNAs may exist at the majority of gene loci [76].

NATs (natural antisense transcripts) are RNA transcripts which overlap a sense protein-coding gene and often act to regulate the associated loci through the recruitment of histone-modifying complexes and induction of transcriptional silencing. Targeting these NATs with siRNAs or antisense oligonucleotides (also known as antagoNATs [88]) results in loss of this epigenetic silencing and consequently, TGA of the sense gene [82,84,85,88] (Figure 6). It has since been demonstrated that miRNAs can also act to regulate transcription by TGS and TGA, suggesting that cross-talk between short and long ncRNA activities is an endogenous mechanism of gene regulation [89–91]. As a result, the use of anti-miRNA oligonucleotides may also be able to disrupt endogenous networks of epigenetic regulation.

Small RNA-mediated transcriptional regulation permits both silencing and activation of therapeutic target genes. Targeting promoter proximal transcripts or NATs allows for gene-specific epigenetic manipulation with numerous potential therapeutic applications. The majority of transcriptional modulation studies have focused on the silencing of the HIV-1 provirus [81,92–94], silencing of oncogenes [95–97] and the activation of tumour suppressors [82,85], although theoretically any gene can be targeted. Recent studies targeting VEGF (vascular endothelial growth factor) [98] and BDNF (brain-derived neurotrophic factor) [99] have reported TGS and TGA *in vivo*, suggesting that these are plausible therapeutic strategies. The ability to regulate genes at the epigenetic level has several advantages over conventional RNAi and antisense oligonucleotide approaches (which regulate gene expression at the post-transcriptional level). Epigenetic modifications are stable and heritable, meaning that long-term silencing can be achieved through a single treatment (or short course of treatments) [86,94,100]. Repeat administration is not required as the silencing or activation effects are not dependent on the presence of effector molecules. Consequently, off-target effects are minimized, saturation of endogenous RNA processing pathways is avoided and the material cost of treatment is greatly reduced.

Conclusions

The use of oligonucleotides to influence miRNA and lncRNA activity is an exciting prospect for future gene therapies. By inhibiting or mimicking miRNA function whole gene networks can be exogenously regulated, providing the potential to attenuate pathological processes. There are numerous instances of miRNA involvement in human pathologies and many of these are attractive therapeutic targets. Given that the target sequence for an anti-miRNA

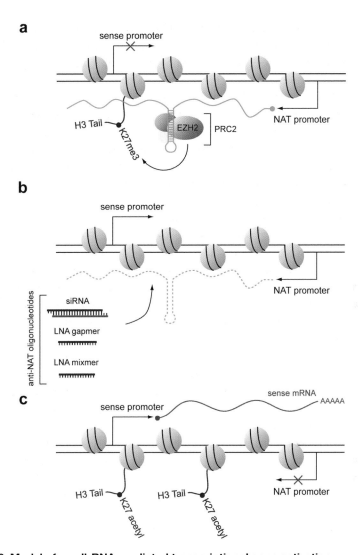

Figure 6. Model of small-RNA-mediated transcriptional gene activation
(**a**) A hypothetical gene locus in which a NAT suppresses transcription of its corresponding sense mRNA by recruiting PRC2. This complex includes the histone methyltransferase EZH2 which tri–methylates Lys^9 on histone H3 in order to induce epigenetic silencing. (**b**) Inhibition of the NAT by siRNAs or antisense oligonucleotides (also known as antagoNATs) leads to abrogation of endogenous epigenetic silencing and loss of repressive histone modifications. (**c**) Sense transcription increases.

oligonucleotide is highly limited, the pool of potential lead compounds to screen is small, in stark contrast with other small molecule drugs. Consequently, it is also practical to screen these compounds *in vivo*. Ongoing progress in the development of chemically modified oligonucleotides will probably lead to nucleic acid drugs with improved pharmacokinetic and pharmacodynamic properties. Indeed, antisense oligonucleotide therapies are steadily advancing towards the clinic and an anti-miRNA therapy for HCV infection is just over the horizon. Additionally, an understanding of miRNA biology is enabling new approaches to enhance existing therapeutic strategies by restricting expression of exogenous transgene expression to specific cell types or by maximizing expression of an endogenous therapeutic gene through abrogation of miRNA regulation.

Although less well understood than miRNAs, it is clear that lncRNAs are highly important in the epigenetic control of gene expression and are potential therapeutic targets for conventional antisense technologies. Specifically, in cases where a lncRNA is directly linked to disease pathogenesis, conventional RNAi or antisense oligonucleotides can be used to modulate its expression. At present, the number of such cases is relatively small, although further investigation into the role that lncRNAs play in disease is likely to identify a plethora of novel therapeutic targets. Given the infancy of this field, the limitations of such approaches are currently unknown. For example, as many lncRNAs exhibit high degrees of secondary structure or are exclusively nuclear-localized, they may be relatively inaccessible to siRNA or antisense oligonucleotide approaches. The use of RNA aptamers targeting lncRNAs could potentially circumvent this problem [101]. Similarly, the complex tissue-specific patterns of expression of some lncRNAs may be an additional obstacle to their therapeutic modulation.

Exogenous small RNAs can regulate gene expression at the transcriptional level by either recruiting chromatin-modifying activities to specific gene promoters and inducing transcriptional silencing, or by disrupting endogenous ncRNA-mediated epigenetic regulation so as to activate transcription. Reciprocal transcriptional silencing and activation, of potentially any gene, can be achieved by targeting a sense pRNA (promoter-associated RNA) or NAT respectively. Initial reports demonstrating *in vivo* epigenetic modulation are highly encouraging [98,99], although future work must address the universal applicability of these phenomena and demonstrate that they are feasible therapeutic strategies.

Summary

- Non-coding RNAs function to regulate gene expression in mammalian cells and are implicated in physiological and pathophysiological processes.
- Oligonucleotides that inhibit or mimic miRNA or long non-coding RNA activity are potential pharmaceutical agents.
- Delivery of oligonucleotides to target tissues remains a substantial obstacle to effective therapy.
- Epigenetic control of gene expression (both silencing and activation) may be possible through the targeting of long non-coding RNAs.

References

1. Clark, M.B., Amaral, P.P., Schlesinger, F.J., Dinger, M.E., Taft, R.J., Rinn, J.L., Ponting, C.P., Stadler, P.F., Morris, K.V., Morillon, A. et al. (2011) The reality of pervasive transcription. PLoS Biol. **9**, e1000625
2. Kapranov, P., Drenkow, J., Cheng, J., Long, J., Helt, G., Dike, S. and Gingeras, T.R. (2005) Examples of the complex architecture of the human transcriptome revealed by RACE and high-density tiling arrays. Genome Res. **15**, 987–997
3. Dahary, D., Elroy-Stein, O. and Sorek, R. (2005) Naturally occurring antisense: transcriptional leakage or real overlap? Genome Res. **15**, 364–368
4. Lapidot, M. and Pilpel, Y. (2006) Genome-wide natural antisense transcription: coupling its regulation to its different regulatory mechanisms. EMBO Rep. **7**, 1216–1222

5. Katayama, S., Tomaru, Y., Kasukawa, T., Waki, K., Nakanishi, M., Nakamura, M., Nishida, H., Yap, C.C., Suzuki, M., Kawai, J. et al. (2005) Antisense transcription in the mammalian transcriptome. Science **309**, 1564–1566
6. Filipowicz, W., Jaskiewicz, L., Kolb, F.A. and Pillai, R.S. (2005) Post-transcriptional gene silencing by siRNAs and miRNAs. Curr. Opin. Struct. Biol. **15**: 331–341
7. Liu, J., Carmell, M.A., Rivas, F.V., Marsden, C.G., Thomson, J.M., Song, J.-J., Hammond, S.M., Joshua-Tor, L. and Hannon, G.J. (2004) Argonaute2 is the catalytic engine of mammalian RNAi. Science **305**, 1437–1441
8. Olsen, P.H. and Ambros, V. (1999) The lin-4 regulatory RNA controls developmental timing in *Caenorhabditis elegans* by blocking LIN-14 protein synthesis after the initiation of translation. Dev. Biol. **216**, 671–680
9. Doench, J.G. and Sharp, P.A. (2004) Specificity of microRNA target selection in translational repression. Genes Dev. **18**, 504–511
10. Ohler, U., Yekta, S., Lim, L.P., Bartel, D.P. and Burge, C.B. (2004) Patterns of flanking sequence conservation and a characteristic upstream motif for microRNA gene identification. RNA **10**, 1309–1322
11. Grimson, A., Farh, K.K.-H., Johnston, W.K., Garrett-Engele, P., Lim, L.P. and Bartel, D.P. (2007) MicroRNA targeting specificity in mammals: determinants beyond seed pairing. Mol. Cell **27**, 91–105
12. Jiang, Q., Wang, Y., Hao, Y., Juan, L., Teng, M., Zhang, X., Li, M., Wang, G. and Liu, Y. (2009) miR2Disease: a manually curated database for microRNA deregulation in human disease. Nucleic Acids Res. **37**, D98–D104
13. Pfeffer, S., Zavolan, M., Grässer, F.A., Chien, M., Russo, J.J., Ju, J., John, B., Enright, A.J., Marks, D., Sander, C. and Tuschl, T. (2004) Identification of virus-encoded microRNAs. Science **304**, 734–736
14. Jopling, C.L., Yi, M., Lancaster, A.M., Lemon, S.M. and Sarnow, P. (2005) Modulation of hepatitis C virus RNA abundance by a liver-specific MicroRNA. Science **309**, 1577–1581
15. Esquela-Kerscher, A. and Slack, F.J. (2006) Oncomirs: microRNAs with a role in cancer. Nat. Rev. Cancer **6**, 259–269
16. Ma, L., Reinhardt, F., Pan, E., Soutschek, J., Bhat, B., Marcusson, E.G., Teruya-Feldstein, J., Bell, G.W. and Weinberg, R.A. (2010) Therapeutic silencing of miR-10b inhibits metastasis in a mouse mammary tumor model. Nat. Biotechnol. **28**, 341–347
17. Choi, W.-Y., Giraldez, A.J. and Schier, A.F. (2007) Target protectors reveal dampening and balancing of Nodal agonist and antagonist by miR-430. Science **318**, 271–274
18. Veedu, R.N. and Wengel, J. (2009) Locked nucleic acid as a novel class of therapeutic agents. RNA Biol. **6**, 321–323
19. Fabani, M.M. and Gait, M.J. (2008) miR-122 targeting with LNA/2′-O-methyl oligonucleotide mixmers, peptide nucleic acids (PNA), and PNA-peptide conjugates. RNA **14**, 336–346
20. Fabani, M.M., Abreu-Goodger, C., Williams, D., Lyons, P.A., Torres, A.G., Smith, K.G.C., Enright, A.J., Gait, M.J. and Vigorito, E. (2010) Efficient inhibition of miR-155 function *in vivo* by peptide nucleic acids. Nucleic Acids Res. **38**, 4466–4475
21. Krützfeldt, J., Rajewsky, N., Braich, R., Rajeev, K.G., Tuschl, T., Manoharan, M. and Stoffel, M. (2005) Silencing of microRNAs *in vivo* with 'antagomirs'. Nature **438**, 685–689
22. Krützfeldt, J., Kuwajima, S., Braich, R., Rajeev, K.G., Pena, J., Tuschl, T., Manoharan, M. and Stoffel, M. (2007) Specificity, duplex degradation and subcellular localization of antagomirs. Nucleic Acids Res. **35**, 2885–2892
23. Esau, C.C. (2008) Inhibition of microRNA with antisense oligonucleotides. Methods **44**, 55–60
24. Alvarez-Erviti, L., Seow, Y., Yin, H., Betts, C., Lakhal, S. and Wood, M.J.A. (2011) Delivery of siRNA to the mouse brain by systemic injection of targeted exosomes. Nat. Biotechnol. **29**, 341–345
25. Lanford, R.E., Hildebrandt-Eriksen, E.S., Petri, A., Persson, R., Lindow, M., Munk, M.E., Kauppinen, S. and Ørum, H. (2010) Therapeutic silencing of microRNA-122 in primates with chronic hepatitis C virus infection. Science **327**, 198–201

26. Ebert, M.S., Neilson, J.R. and Sharp, P.A. (2007) MicroRNA sponges: competitive inhibitors of small RNAs in mammalian cells. Nat. Methods **4**, 721–726
27. Ma, L., Young, J., Prabhala, H., Pan, E., Mestdagh, P., Muth, D., Teruya-Feldstein, J., Reinhardt, F., Onder, T.T., Valastyan, S. et al. (2010) miR-9, a MYC/MYCN-activated microRNA, regulates E-cadherin and cancer metastasis. Nat. Cell Biol. **12**, 247–256
28. Kay, M.A. (2011) State-of-the-art gene-based therapies: the road ahead. Nat. Rev. Genet. **12**, 316–328
29. Cacchiarelli, D., Incitti, T., Martone, J., Cesana, M., Cazzella, V., Santini, T., Sthandier, O. and Bozzoni, I. (2011) miR-31 modulates dystrophin expression: new implications for Duchenne muscular dystrophy therapy. EMBO Rep. **12**, 136–141
30. Calin, G.A. and Croce, C.M. (2006) MicroRNA signatures in human cancers. Nat. Rev. Cancer **6**, 857–866
31. Hammond, S.M. (2007) MicroRNAs as tumor suppressors. Nat. Genet. **39**, 582–583
32. Kota, J., Chivukula, R.R., O'Donnell, K.A., Wentzel, E.A., Montgomery, C.L., Hwang, H.-W., Chang, T.-C., Vivekanandan, P., Torbenson, M., Clark, K.R. et al. (2009) Therapeutic microRNA delivery suppresses tumorigenesis in a murine liver cancer model. Cell **137**, 1005–1017
33. van Rooij, E., Sutherland, L.B., Thatcher, J.E., DiMaio, J.M., Naseem, R.H., Marshall, W.S., Hill, J.A. and Olson, E.N. (2008) Dysregulation of microRNAs after myocardial infarction reveals a role of miR-29 in cardiac fibrosis. Proc. Natl. Acad. Sci. U.S.A. **105**, 13027–13032
34. Wang, L., Zhou, L., Jiang, P., Lu, L., Chen, X., Lan, H., Guttridge, D.C., Sun, H. and Wang, H. (2012) Loss of miR-29 in myoblasts contributes to dystrophic muscle pathogenesis. Mol. Ther. **20**, 1222–1233
35. Rettig, G.R. and Behlke, M.A. (2012) Progress toward *in vivo* use of siRNAs-II. Mol. Ther. **20**, 483–512
36. Brown, B.D., Venneri, M.A., Zingale, A., Sergi Sergi, L. and Naldini, L. (2006) Endogenous microRNA regulation suppresses transgene expression in hematopoietic lineages and enables stable gene transfer. Nat. Med. **12**, 585–591
37. Cawood, R., Chen, H.H., Carroll, F., Bazan-Peregrino, M., van Rooijen, N. and Seymour, L.W. (2009) Use of tissue-specific microRNA to control pathology of wild-type adenovirus without attenuation of its ability to kill cancer cells. PLoS Pathog. **5**, e1000440
38. Mercer, T.R., Dinger, M.E., Sunkin, S.M., Mehler, M.F. and Mattick, J.S. (2008) Specific expression of long noncoding RNAs in the mouse brain. Proc. Natl. Acad. Sci. U.S.A. **105**, 716–721
39. Dinger, M.E., Amaral, P.P., Mercer, T.R., Pang, K.C., Bruce, S.J., Gardiner, B.B., Askarian-Amiri, M.E., Ru, K., Soldà, G., Simons, C. et al. (2008) Long noncoding RNAs in mouse embryonic stem cell pluripotency and differentiation. Genome Res. **18**, 1433–1445
40. Guttman, M., Amit, I., Garber, M., French, C., Lin, M.F., Feldser, D., Huarte, M., Zuk, O., Carey, B.W., Cassady, J.P. et al. (2009) Chromatin signature reveals over a thousand highly conserved large non-coding RNAs in mammals. Nature **458**, 223–227
41. Bonasio, R., Tu, S. and Reinberg, D. (2010) Molecular signals of epigenetic states. Science **330**, 612–616
42. Sun, B.K., Deaton, A.M. and Lee, J.T. (2006) A transient heterochromatic state in Xist preempts X inactivation choice without RNA stabilization. Mol. Cell **21**, 617–628
43. Mayer, C., Schmitz, K.-M., Li, J., Grummt, I. and Santoro, R. (2006) Intergenic transcripts regulate the epigenetic state of rRNA genes. Mol. Cell **22**, 351–361
44. Schmitz, K.-M., Mayer, C., Postepska, A. and Grummt, I. (2010) Interaction of noncoding RNA with the rDNA promoter mediates recruitment of DNMT3b and silencing of rRNA genes. Genes Dev. **24**, 2264–2269
45. Huarte, M., Guttman, M., Feldser, D., Garber, M., Koziol, M.J., Kenzelmann-Broz, D., Khalil, A.M., Zuk, O., Amit, I., Rabani, M. et al. (2010) A large intergenic noncoding RNA induced by p53 mediates global gene repression in the p53 response. Cell **142**, 409–419
46. Rinn, J.L., Kertesz, M., Wang, J.K., Squazzo, S.L., Xu, X., Brugmann, S.A., Goodnough, L.H., Helms, J.A., Farnham, P.J., Segal, E. and Chang, H.Y. (2007) Functional demarcation of active and silent chromatin domains in human HOX loci by noncoding RNAs. Cell **129**, 1311–1323

47. Tsai, M.-C., Manor, O., Wan, Y., Mosammaparast, N., Wang, J.K., Lan, F., Shi, Y., Segal, E. and Chang, H.Y. (2010) Long noncoding RNA as modular scaffold of histone modification complexes. Science **329**, 689–693
48. Maison, C., Bailly, D., Roche, D., Montes de Oca, R., Probst, A.V., Vassias, I., Dingli, F., Lombard, B., Loew, D., Quivy, J.-P. and Almouzni, G. (2011) SUMOylation promotes de novo targeting of HP1α to pericentric heterochromatin. Nat. Genet. **43**, 220–227
49. Khalil, A.M., Guttman, M., Huarte, M., Garber, M., Raj, A., Rivea Morales, D., Thomas, K., Presser, A., Bernstein, B.E., van Oudenaarden, A. et al. (2009) Many human large intergenic noncoding RNAs associate with chromatin-modifying complexes and affect gene expression. Proc. Natl. Acad. Sci. U.S.A. **106**, 11667–11672
50. Wang, K.C., Yang, Y.W., Liu, B., Sanyal, A., Corces-Zimmerman, R., Chen, Y., Lajoie, B.R., Protacio, A., Flynn, R.A., Gupta, R.A. et al. (2011) A long noncoding RNA maintains active chromatin to coordinate homeotic gene expression. Nature **472**, 120–124
51. Yang, L., Lin, C., Liu, W., Zhang, J., Ohgi, K.A., Grinstein, J.D., Dorrestein, P.C. and Rosenfeld, M.G. (2011) ncRNA- and Pc2 methylation-dependent gene relocation between nuclear structures mediates gene activation programs. Cell **147**, 773–788
52. Kino, T., Hurt, D.E., Ichijo, T., Nader, N. and Chrousos, G.P. (2010) Noncoding RNA gas5 is a growth arrest- and starvation-associated repressor of the glucocorticoid receptor. Sci. Signaling **3**, ra8
53. Cesana, M., Cacchiarelli, D., Legnini, I., Santini, T., Sthandier, O., Chinappi, M., Tramontano, A. and Bozzoni, I. (2011) A long noncoding RNA controls muscle differentiation by functioning as a competing endogenous RNA. Cell **147**, 358–369
54. Poliseno, L., Salmena, L., Zhang, J., Carver, B., Haveman, W.J. and Pandolfi, P.P. (2010) A coding-independent function of gene and pseudogene mRNAs regulates tumour biology. Nature **465**, 1033–1038
55. Faghihi, M.A., Zhang, M., Huang, J., Modarresi, F., Van der Brug, M.P., Nalls, M.A., Cookson, M.R., St-Laurent, 3rd, G. and Wahlestedt, C. (2010) Evidence for natural antisense transcript-mediated inhibition of microRNA function. Genome Biol. **11**, R56
56. Bernard, D., Prasanth, K.V., Tripathi, V., Colasse, S., Nakamura, T., Xuan, Z., Zhang, M.Q., Sedel, F., Jourdren, L., Coulpier, F. et al. (2010) A long nuclear-retained non-coding RNA regulates synaptogenesis by modulating gene expression. EMBO J. **29**, 3082–3093
57. Tripathi, V., Ellis, J.D., Shen, Z., Song, D.Y., Pan, Q., Watt, A.T., Freier, S.M., Bennett, C.F., Sharma, A., Bubulya, P.A. et al. (2010) The nuclear-retained noncoding RNA MALAT1 regulates alternative splicing by modulating SR splicing factor phosphorylation. Mol. Cell **39**, 925–938
58. Petruk, S., Sedkov, Y., Riley, K.M., Hodgson, J., Schweisguth, F., Hirose, S., Jaynes, J.B., Brock, H.W. and Mazo, A. (2006) Transcription of bxd noncoding RNAs promoted by trithorax represses Ubx in cis by transcriptional interference. Cell **127**, 1209–1221
59. Martianov, I., Ramadass, A., Serra Barros, A., Chow, N. and Akoulitchev, A. (2007) Repression of the human dihydrofolate reductase gene by a non-coding interfering transcript. Nature **445**, 666–670
60. Faghihi, M.A., Modarresi, F., Khalil, A.M., Wood, D.E., Sahagan, B.G., Morgan, T.E., Finch, C.E., St Laurent, 3rd, G., Kenny, P.J. and Wahlestedt, C. (2008) Expression of a noncoding RNA is elevated in Alzheimer's disease and drives rapid feed–forward regulation of β–secretase. Nat. Med. **14**, 723–730
61. Modarresi, F., Faghihi, M.A., Patel, N.S., Sahagan, B.G., Wahlestedt, C. and Lopez-Toledano, M.A. (2011) Knockdown of BACE1-AS nonprotein-coding transcript modulates β-amyloid-related hippocampal neurogenesis. Int. J. Alzheimers Dis. **2011**, 929042
62. Yang, N. and Kazazian, Jr, H.H. (2006) L1 retrotransposition is suppressed by endogenously encoded small interfering RNAs in human cultured cells. Nat. Struct. Mol. Biol. **13**, 763–771
63. Kawaji, H., Nakamura, M., Takahashi, Y., Sandelin, A., Katayama, S., Fukuda, S., Daub, C.O., Kai, C., Kawai, J., Yasuda, J. et al. (2008) Hidden layers of human small RNAs. BMC Genomics **9**, 157

64. Qureshi, I.A., Mattick, J.S. and Mehler, M.F. (2010) Long non-coding RNAs in nervous system function and disease. Brain Res. **1338**, 20–35
65. Guttman, M., Donaghey, J., Carey, B.W., Garber, M., Grenier, J.K., Munson, G., Young, G., Lucas, A.B., Ach, R., Bruhn, L. et al. (2011) lincRNAs act in the circuitry controlling pluripotency and differentiation. Nature **477**, 295–300
66. Panning, B. and Jaenisch, R. (1998) RNA and the epigenetic regulation of X chromosome inactivation. Cell **93**, 305–308
67. Rougeulle, C. and Heard, E. (2002) Antisense RNA in imprinting: spreading silence through Air. Trends Genet. **18**, 434–437
68. Halvorsen, M., Martin, J.S., Broadaway, S. and Laederach, A. (2010) Disease-associated mutations that alter the RNA structural ensemble. PLoS Genet. **6**: e1001074
69. Glinskii, A.B., Ma, J., Ma, S., Grant, D., Lim, C.-U., Sell, S. and Glinsky, G.V. (2009) Identification of intergenic trans-regulatory RNAs containing a disease-linked SNP sequence and targeting cell cycle progression/differentiation pathways in multiple common human disorders. Cell Cycle **8**, 3925–3942
70. Wojcik, S.E., Rossi, S., Shimizu, M., Nicoloso, M.S., Cimmino, A., Alder, H., Herlea, V., Rassenti, L.Z., Rai, K.R., Kipps, T.J. et al. (2010) Non-codingRNA sequence variations in human chronic lymphocytic leukemia and colorectal cancer. Carcinogenesis **31**, 208–215
71. Gupta, R.A., Shah, N., Wang, K.C., Kim, J., Horlings, H.M., Wong, D.J., Tsai, M.-C., Hung, T., Argani, P., Rinn, J.L. et al. (2010) Long non-coding RNA HOTAIR reprograms chromatin state to promote cancer metastasis. Nature **464**, 1071–1076
72. Ji, P., Diederichs, S., Wang, W., Böing, S., Metzger, R., Schneider, P.M., Tidow, N., Brandt, B., Buerger, H., Bulk, E. et al. (2003) MALAT-1, a novel noncoding RNA, and thymosin β4 predict metastasis and survival in early-stage non-small cell lung cancer. Oncogene **22**, 8031–8041
73. Tufarelli, C., Stanley, J.A.S., Garrick, D., Sharpe, J.A., Ayyub, H., Wood, W.G. and Higgs, D.R. (2003) Transcription of antisense RNA leading to gene silencing and methylation as a novel cause of human genetic disease. Nat. Genet. **34**, 157–165
74. Lewejohann, L., Skryabin, B.V., Sachser, N., Prehn, C., Heiduschka, P., Thanos, S., Jordan, U., Dell'Omo, G., Vyssotski, A.L., Pleskacheva, M.G. et al. (2004) Role of a neuronal small non-messenger RNA: behavioural alterations in BC1 RNA-deleted mice. Behav. Brain Res. **154**, 273–289
75. Mus, E., Hof, P.R. and Tiedge, H. (2007) Dendritic BC200 RNA in aging and in Alzheimer's disease. Proc. Natl. Acad. Sci. U.S.A. **104**, 10679–10684
76. Preker, P., Nielsen, J., Kammler, S., Lykke-Andersen, S., Christensen, M.S., Mapendano, C.K., Schierup, M.H. and Jensen, T.H. (2008) RNA exosome depletion reveals transcription upstream of active human promoters. Science **322**, 1851–1854
77. Ponjavic, J. and Ponting, C.P. (2007) The long and the short of RNA maps. BioEssays **29**, 1077–1080
78. Ørom, U.A. and Shiekhattar, R. (2011) Long non-coding RNAs and enhancers. Curr. Opin. Genet. Dev. **21**, 194–198
79. Morris, K.V., Chan, S.W.-L., Jacobsen, S.E. and Looney, D.J. (2004) Small interfering RNA-induced transcriptional gene silencing in human cells. Science **305**, 1289–1292
80. Han, J., Kim, D. and Morris, K.V. (2007) Promoter-associated RNA is required for RNA-directed transcriptional gene silencing in human cells. Proc. Natl. Acad. Sci. U.S.A **104**, 12422–12427
81. Weinberg, M.S., Villeneuve, L.M., Ehsani, A., Amarzguioui, M., Aagaard, L., Chen, Z.-X., Riggs, A.D., Rossi, J.J. and Morris, K.V. (2006) The antisense strand of small interfering RNAs directs histone methylation and transcriptional gene silencing in human cells. RNA **12**, 256–262
82. Li, L.-C., Okino, S.T., Zhao, H., Pookot, D., Place, R.F., Urakami, S., Enokida, H. and Dahiya, R. (2006) Small dsRNAs induce transcriptional activation in human cells. Proc. Natl. Acad. Sci. U.S.A **103**, 17337–17342
83. Janowski, B.A., Younger, S.T., Hardy, D.B., Ram, R., Huffman, K.E. and Corey, D.R. (2007) Activating gene expression in mammalian cells with promoter-targeted duplex RNAs. Nat. Chem. Biol. **3**, 166–173

84. Schwartz, J.C., Younger, S.T., Nguyen, N.-B., Hardy, D.B., Monia, B.P., Corey, D.R. and Janowski, B.A. (2008) Antisense transcripts are targets for activating small RNAs. Nat. Struct. Mol. Biol. **15**, 842–848
85. Morris, K.V., Santoso, S., Turner, A.-M., Pastori, C. and Hawkins, P.G. (2008) Bidirectional transcription directs both transcriptional gene activation and suppression in human cells. PLoS Genet. **4**, e1000258
86. Hawkins, P.G., Santoso, S., Adams, C., Anest, V. and Morris, K.V. (2009) Promoter targeted small RNAs induce long-term transcriptional gene silencing in human cells. Nucleic Acids Res. **37**, 2984–2995
87. Kim, D.H., Villeneuve, L.M., Morris, K.V. and Rossi, J.J. (2006) Argonaute-1 directs siRNA-mediated transcriptional gene silencing in human cells. Nat. Struct. Mol. Biol. **13**, 793–797
88. Modarresi, F., Faghihi, M.A., Lopez-Toledano, M.A., Fatemi, R.P., Magistri, M., Brothers, S.P., van der Brug, M.P. and Wahlestedt, C. (2012) Inhibition of natural antisense transcripts in vivo results in gene-specific transcriptional upregulation. Nat. Biotechnol. **30**, 453–459
89. Kim, D.H., Saetrom, P., Snøve, O. and Rossi, J.J. (2008) MicroRNA-directed transcriptional gene silencing in mammalian cells. Proc. Natl. Acad. Sci. U.S.A. **105**, 16230–16235
90. Place, R.F., Li, L.-C., Pookot, D., Noonan, E.J. and Dahiya, R. (2008) MicroRNA-373 induces expression of genes with complementary promoter sequences. Proc. Natl. Acad. Sci. U.S.A. **105**, 1608–1613
91. Younger, S.T., Pertsemlidis, A. and Corey, D.R. (2009) Predicting potential miRNA target sites within gene promoters. Bioorg. Med. Chem. Lett. **19**, 3791–3794
92. Suzuki, K., Shijuuku, T., Fukamachi, T., Zaunders, J., Guillemin, G., Cooper, D. and Kelleher, A. (2005) Prolonged transcriptional silencing and CpG methylation induced by siRNAs targeted to the HIV-1 promoter region. J RNAi Gene Silencing **1**, 66–78
93. Suzuki, K., Juelich, T., Lim, H., Ishida, T., Watanebe, T., Cooper, D.A., Rao, S. and Kelleher, A.D. (2008) Closed chromatin architecture is induced by an RNA duplex targeting the HIV-1 promoter region. J. Biol. Chem. **283**, 23353–23363
94. Turner, A.-M.W., De La Cruz, J. and Morris, K.V. (2009) Mobilization-competent lentiviral vector-mediated sustained transcriptional modulation of HIV-1 expression. Mol. Ther. **17**, 360–368
95. Pulukuri, S.M.K. and Rao, J.S. (2007) Small interfering RNA directed reversal of urokinase plasminogen activator demethylation inhibits prostate tumor growth and metastasis. Cancer Res. **67**, 6637–6646
96. Ting, A.H., Schuebel, K.E., Herman, J.G. and Baylin, S.B. (2005) Short double-stranded RNA induces transcriptional gene silencing in human cancer cells in the absence of DNA methylation. Nat Genet. **37**, 906–910
97. Mehndiratta, M., Palanichamy, J.K., Pal, A., Bhagat, M., Singh, A., Sinha, S. and Chattopadhyay, P. (2011) CpG hypermethylation of the C-myc promoter by dsRNA results in growth suppression. Mol. Pharmacol. **8**, 2302–2309
98. Turunen, M.P., Lehtola, T., Heinonen, S.E., Assefa, G.S., Korpisalo, P., Girnary, R., Glass, C.K., Väisänen, S. and Ylä-Herttuala, S. (2009) Efficient regulation of VEGF expression by promoter-targeted lentiviral shRNAs based on epigenetic mechanism: a novel example of epigenetherapy. Circ. Res. **105**, 604–609
99. Modarresi, F., Faghihi, M.A., Lopez-Toledano, M.A., Fatemi, R.P., Magistri, M., Brothers, S.P., van der Brug, M.P. and Wahlestedt, C. (2012) Inhibition of natural antisense transcripts *in vivo* results in gene-specific transcriptional upregulation. Nat. Biotechnol. **30**, 453–459
100. Yamagishi, M., Ishida, T., Miyake, A., Cooper, D.A., Kelleher, A.D., Suzuki, K. and Watanabe, T. (2009) Retroviral delivery of promoter-targeted shRNA induces long-term silencing of HIV–1 transcription. Microbes Infect. **11**, 500–508
101. Tsai, M.-C., Spitale, R.C. and Chang, H.Y. (2011) Long intergenic noncoding RNAs: new links in cancer progression. Cancer Res. **71**, 3–7

INDEX

A

acute lymphoblastic leukaemia, 121
ADAR, 19, 95
adenosine deaminases that act on RNA (see ADAR)
adenovirus, 24, 132
Airn, 7, 97, 114, 118, 119
AGO/Ago (see also Argonaute), 5, 6, 22, 24, 32, 34, 35, 40, 41, 42, 43, 44, 45, 46, 48, 98, 128, 129, 137, 138
Alzheimer's disease, 7, 95, 121, 135
AMO, 130, 132
Amphimedon queenslandica, 8
ANRIL, 119, 121–122
antagomir, 129, 130, 131
antagoNAT (see also natural antisense transcript), 97, 138, 139
anti-mRNA oligonucleotide (see AMO)
antisense, 3, 7, 8, 42, 43, 44, 55, 65, 91–99, 106, 107, 114, 119, 121, 122, 127, 128, 130, 133, 135, 136, 137, 138, 139, 140
aortic aneurysm, 121
Arabidopsis thaliana, 33, 36, 66
Argonaute, 4, 6, 17, 18, 20, 21, 25, 39, 40, 45, 59, 96, 98

B

BLV, 24
bovine leukaemia virus (see BLV)

C

Caenorhabditis elegans, 4, 8, 18, 20, 21, 31, 62
cancer, 6, 20, 48, 49, 53, 54, 55, 105, 106, 108, 113, 119, 121, 122, 130, 132, 135
catalysis, 82, 84, 85, 86, 88
C/D box RNA, 54, 55, 56, 57, 58, 59, 60, 61, 62, 63, 64, 65, 66, 67, 68, 69, 70
CDKN2B anisense RNA 1 (see ANRIL)
chronic lymphocytic leukaemia (see CLL)
cis-regulation, 113, 120
cleavage and polyadenylation-stimulating factor (see CPSF)

CLL, 135
clustered regularly interspersed short palindromic repeats (see CRISPR)
colorectal cancer (see CRC)
coronary artery disease, 121
CPSF, 87
CRC, 135
CRISPR, 8
CRISPR RNA (see crRNA)
crRNA, 8

D

deadenylation, 32, 33, 36, 41
decapping, 31, 32, 36
degradome, 29, 34, 36
development, 1, 2, 9, 10, 17, 18, 33, 40, 41, 48, 49, 54, 96, 99, 105, 114
DGCR8, 18, 19, 22, 129
diabetes, 60, 105, 108, 121
Dicer, 17, 18, 19, 20, 21, 22, 24, 59, 95, 96, 106, 128, 129, 135, 136
DiGeorge syndrome critical region gene 8 (see DGCR8)
DMD, 130, 132
DNA methylation, 6, 40, 41, 95, 134, 137, 138
dosage compensation, 93, 113, 118
double-stranded DNA (see dsDNA)
double-stranded RNA (see dsRNA)
Drosha, 17, 18, 19, 20, 21, 22, 24, 95, 128, 129
Drosophila melanogaster, 21, 39, 40, 41, 42, 44, 45, 46, 48, 114, 117, 118, 119, 121
Drosophilid, 121
dsDNA, 24, 35
dsRNA, 4, 18, 19, 20, 21, 95, 96, 106, 107, 135
Duchenne muscular dystrophy (see DMD)

E

EBV, 24
ENCODE project, 3, 116
endogenous siRNA (see endo-siRNA)
endo-siRNA, 96, 98, 99, 135, 136
enhancer RNA (see eRNA)

epigenetics, 6, 9, 92, 96, 97, 98, 99, 127, 133, 134, 135, 136, 138, 139, 140
Epstein–Barr virus (see EBV)
eRNA, 120
Escherichia coli, 8
EST, 115
Euglena gracilis, 57
Evf2, 114, 118, 120
evolutionary conservation, 113, 133
exon, 3, 7, 60, 61, 80, 81, 82, 83, 84, 85, 87, 94, 104, 113, 121, 131
expressed sequence tag (see EST)

G

gene
    expression, 1, 2, 4, 7, 10, 17, 18, 20, 29, 60, 70, 78, 80, 87, 88, 91, 92, 93, 96, 98, 105, 106, 108, 109, 113, 117, 118, 120, 121, 122, 127, 128, 133, 134, 135, 137, 138, 140
    silencing, 4, 40, 41, 59, 91, 92, 98, 127, 134, 136, 137
    therapy, 130, 132
gene termini-associated small RNA (see TASR)
genome, 1, 2, 3, 4, 6, 7, 8, 9, 18, 24, 34, 36, 39, 40, 41, 44, 49, 57, 61, 62, 65, 80, 91, 99, 104, 105, 109, 114, 116, 118, 119, 122, 128, 135, 136
genomic quality control, 98, 99
germline, 39, 40, 41, 42, 43, 44, 48, 49
Giardia lamblia, 59, 62
glioma, 121
group II intron, 79, 80, 84, 85, 86, 88

H

H/ACA box RNA, 23, 24, 54, 55, 56, 57, 58, 59, 60, 61, 63, 65, 66, 67, 69, 70
HCV, 130, 139
Helicobacter pylori, 8
hepatitis C virus (see HCV)
herpes simplex virus, 24
heterochromatin protein 1 (see HP1)
histone modification, 40, 41, 96, 139
HP1, 44, 46, 134

I

ICG, 134
imprinting, 7, 92, 93, 113
interchromatin granule (see ICG)

intergenic, 7, 40, 113, 114, 115, 120, 123, 128, 133
intron (see also group II intron), 3, 7, 18, 21, 23, 61, 62, 79, 80, 81, 82, 83, 84, 87, 94, 104, 128, 133
ischaemic stroke, 121

K

Kaposi's sarcoma-associated herpes virus (see KSHV)
KSHV, 24

L

let-7, 19, 20
lin-4, 31
lincRNA, 7, 114, 115, 116, 117, 118, 119, 120, 121, 122, 123, 133, 134, 136
Listeria monocytogenes, 8
LNA, 130, 131
lncRNA, 5, 6–7, 61, 105, 113, 114, 115, 118, 119, 120, 122, 123, 127, 128, 132, 133, 134, 135, 136, 138, 140
locked nucleic acid (see LNA)
long intergenic non-coding RNA (see lincRNA)
long non-coding RNA (see lncRNA)
Lymnaea stagnalis, 106

M

MALAT-1, 118, 121, 122
malignant melanoma, 121
Marek's disease virus (see MDV)
MDV, 24
metastasis associated in lung adenocarcinoma transcript-1 (see MALAT-1)
Methanosarcina mazei, 8
miRISC, 20
miRLC, 21
miRNA-induced silencing complex (see miRISC)
miRNA loading complex (see miRLC)
miRNA mimic, 127, 129, 132, 140
monoallelic gene expression, 98

N

NAT, 91–99, 138, 139, 140
natural antisense transcript (see NAT)
NFAT, 118, 121
nuclear factor of activated T-cells (see NFAT)